Aquatic Ecology

Aquatic Ecology

Edited by Vincent Jennings

SYRAWOOD
PUBLISHING HOUSE
New York

Published by Syrawood Publishing House,
750 Third Avenue, 9th Floor,
New York, NY 10017, USA
www.syrawoodpublishinghouse.com

Aquatic Ecology
Edited by Vincent Jennings

International Standard Book Number: 978-1-68286-615-3 (Hardback)

Cataloging-in-Publication Data

Aquatic ecology / edited by Vincent Jennings.
p. cm.
Includes bibliographical references and index.
ISBN 978-1-68286-615-3
1. Aquatic ecology. 2. Ecology. 3. Aquatic biology. I. Jennings, Vincent.
QH541.5.W3 A68 2018
577.6--dc23

Printed in China.

TABLE OF CONTENTS

Preface VII

Chapter 1 **An Introduction to Aquatic Ecology and Ecosystems** 1
- Aquatic Ecosystem 1
- Marine Ecosystem 17
- Marine Habitats 18
- Marsh 34

Chapter 2 **An Overview of Aquatic Plants** 41
- Aquatic Plant 41
- Macrophyte 64

Chapter 3 **A Comprehensive Study of Aquatic Animals** 68
- Aquatic Animal 68
- Aquatic Mammal 104
- Baikal Seal 138

Chapter 4 **Marine Invertebrates: An Integrated Study** 160
- Marine Invertebrates 160
- Annelid 192
- Brachiopod 204
- Bryozoa 217
- Chaetognatha 234

Permissions

Index

PREFACE

All the water bodies on Earth have an aquatic ecosystem and an aquatic ecology. It consists of all the under-water plants, animals, etc. Aquatic ecology can be divided into two main parts namely freshwater ecosystem and marine ecosystem. This book is a valuable compilation of topics, ranging from the basic to the most complex discoveries and principles in the field of aquatic ecology. The topics included in it are of utmost significance and bound to provide incredible insights to readers. Those with an interest in the subject of aquatic ecology would find this textbook helpful.

A short introduction to every chapter is written below to provide an overview of the content of the book:

Chapter 1 - Aquatic ecology is the study of the interaction that occurs within aquatic organisms and with their environment. They can be of different types like marine, lotic, wetlands, freshwater, etc. This is an introductory chapter which will introduce briefly all the significant aspects of aquatic ecology and ecosystem; **Chapter 2 -** The plants found in the water bodies are known as aquatic plants. Equisetum fluviatile, hippuris vulgaris, sagittaria, ceratophyllum and seagrass are some of the types of aquatic plants. This chapter is an overview of the subject matter incorporating all the major aspects of aquatic plants; **Chapter 3 -** Aquatic animals live within a water body and breathe the oxygen dissolved in the water. Some of the types of aquatic animals are river dolphin, cavefish, freshwater snail and whale. The topics elaborated in this chapter will help in gaining a better perspective about aquatic animals; **Chapter 4 -** Marine invertebrates are animals without a vertebral column that are found within marine habitats. Some of the marine invertebrates are annelid, brachiopod, bryozoa and chaetognatha. The aspects elucidated in this chapter are of vital importance, and provide a better understanding of marine invertebrates.

Finally, I would like to thank my fellow scholars who gave constructive feedback and my family members who supported me at every step.

Editor

1

An Introduction to Aquatic Ecology and Ecosystems

Aquatic ecology is the study of the interaction that occurs within aquatic organisms and with their environment. They can be of different types like marine, lotic, wetlands, freshwater, etc. This is an introductory chapter which will introduce briefly all the significant aspects of aquatic ecology and ecosystem.

Aquatic Ecosystem

An estuary mouth and coastal waters, part of an aquatic ecosystem

An aquatic ecosystem is an ecosystem in a body of water. Communities of organisms that are dependent on each other and on their environment live in aquatic ecosystems. The two main types of aquatic ecosystems are marine ecosystems and freshwater ecosystems.

Types

Marine

Marine ecosystems cover approximately 71% of the Earth's surface and contain approximately 97% of the planet's water. They generate 32% of the world's net primary production. They are distinguished from freshwater ecosystems by the presence of dissolved compounds, especially salts, in the water. Approximately 85% of the dissolved materials in seawater are sodium and chlorine. Seawater has an average salinity of 35 parts per thousand (ppt) of water. Actual salinity varies among different marine ecosystems.

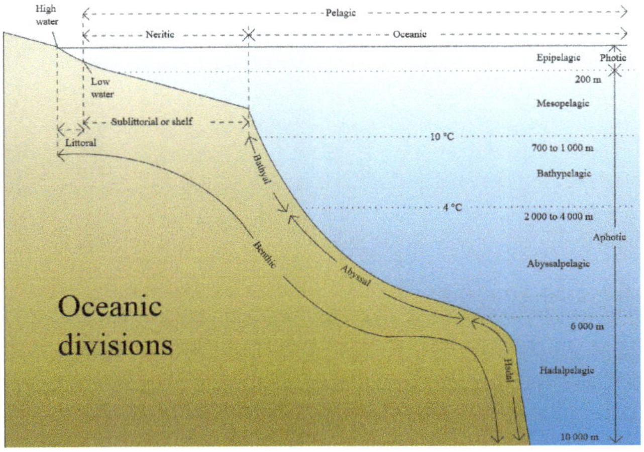

A classification of marine habitats.

Marine ecosystems can be divided into many zones depending upon water depth and shoreline features. The oceanic zone is the vast open part of the ocean where animals such as whales, sharks, and tuna live. The benthic zone consists of substrates below water where many invertebrates live. The intertidal zone is the area between high and low tides; in this figure it is termed the littoral zone. Other near-shore (neritic) zones can include estuaries, salt marshes, coral reefs, lagoons and mangrove swamps. In the deep water, hydrothermal vents may occur where chemosynthetic sulfur bacteria form the base of the food web.

Classes of organisms found in marine ecosystems include brown algae, dinoflagellates, corals, cephalopods, echinoderms, and sharks. Fishes caught in marine ecosystems are the biggest source of commercial foods obtained from wild populations.

Environmental problems concerning marine ecosystems include unsustainable exploitation of marine resources (for example overfishing of certain species), marine pollution, climate change, and building on coastal areas.

Freshwater

Freshwater ecosystem.

Freshwater ecosystems cover 0.78% of the Earth's surface and inhabit 0.009% of its total water. They generate nearly 3% of its net primary production. Freshwater ecosystems contain 41% of the world's known fish species.

There are three basic types of freshwater ecosystems:

- Lentic: slow moving water, including pools, ponds, and lakes.

- Lotic: faster moving water, for example streams and rivers.
- Wetlands: areas where the soil is saturated or inundated for at least part of the time.

Lentic

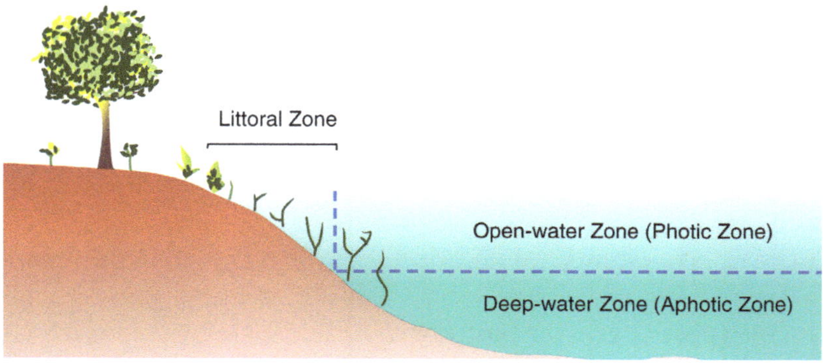

The three primary zones of a lake.

Lake ecosystems can be divided into zones. One common system divides lakes into three zones. The first, the littoral zone, is the shallow zone near the shore. This is where rooted wetland plants occur. The offshore is divided into two further zones, an open water zone and a deep water zone. In the open water zone (or photic zone) sunlight supports photosynthetic algae, and the species that feed upon them. In the deep water zone, sunlight is not available and the food web is based on detritus entering from the littoral and photic zones. Some systems use other names. The off shore areas may be called the pelagic zone, the photic zone may be called the limnetic zone and the aphotic zone may be called the profundal zone. Inland from the littoral zone one can also frequently identify a riparian zone which has plants still affected by the presence of the lake—this can include effects from windfalls, spring flooding, and winter ice damage. The production of the lake as a whole is the result of production from plants growing in the littoral zone, combined with production from plankton growing in the open water.

Wetlands can be part of the lentic system, as they form naturally along most lake shores, the width of the wetland and littoral zone being dependent upon the slope of the shoreline and the amount of natural change in water levels, within and among years. Often dead trees accumulate in this zone, either from windfalls on the shore or logs transported to the site during floods. This woody debris provides important habitat for fish and nesting birds, as well as protecting shorelines from erosion.

Two important subclasses of lakes are ponds, which typically are small lakes that intergrade with wetlands, and water reservoirs. Over long periods of time, lakes, or bays within them, may gradually become enriched by nutrients and slowly fill in with organic sediments, a process called succession. When humans use the watershed, the volumes of sediment entering the lake can accelerate this process. The addition of sediments and nutrients to a lake is known as eutrophication.

Ponds

Ponds are small bodies of freshwater with shallow and still water, marsh, and aquatic plants. They can be further divided into four zones: vegetation zone, open water, bottom mud and surface film.

The size and depth of ponds often varies greatly with the time of year; many ponds are produced by spring flooding from rivers. Food webs are based both on free-floating algae and upon aquatic plants. There is usually a diverse array of aquatic life, with a few examples including algae, snails, fish, beetles, water bugs, frogs, turtles, otters and muskrats. Top predators may include large fish, herons, or alligators. Since fish are a major predator upon amphibian larvae, ponds that dry up each year, thereby killing resident fish, provide important refugia for amphibian breeding. Ponds that dry up completely each year are often known as vernal pools. Some ponds are produced by animal activity, including alligator holes and beaver ponds, and these add important diversity to landscapes.

Lotic

The major zones in river ecosystems are determined by the river bed's gradient or by the velocity of the current. Faster moving turbulent water typically contains greater concentrations of dissolved oxygen, which supports greater biodiversity than the slow moving water of pools. These distinctions form the basis for the division of rivers into upland and lowland rivers. The food base of streams within riparian forests is mostly derived from the trees, but wider streams and those that lack a canopy derive the majority of their food base from algae. Anadromous fish are also an important source of nutrients. Environmental threats to rivers include loss of water, dams, chemical pollution and introduced species. A dam produces negative effects that continue down the watershed. The most important negative effects are the reduction of spring flooding, which damages wetlands, and the retention of sediment, which leads to loss of deltaic wetlands.

Wetlands

Wetlands are dominated by vascular plants that have adapted to saturated soil. There are four main types of wetlands: swamp, marsh, fen and bog (both fens and bogs are types of mire). Wetlands are the most productive natural ecosystems in the world because of the proximity of water and soil. Hence they support large numbers of plant and animal species. Due to their productivity, wetlands are often converted into dry land with dykes and drains and used for agricultural purposes. The construction of dykes, and dams, has negative consequences for individual wetlands and entire watersheds. Their closeness to lakes and rivers means that they are often developed for human settlement. Once settlements are constructed and protected by dykes, the settlements then become vulnerable to land subsidence and ever increasing risk of flooding. The Louisiana coast around New Orleans is a well-known example; the Danube Delta in Europe is another.

Functions

Aquatic ecosystems perform many important environmental functions. For example, they recycle nutrients, purify water, attenuate floods, recharge ground water and provide habitats for wildlife. Aquatic ecosystems are also used for human recreation, and are very important to the tourism industry, especially in coastal regions.

The health of an aquatic ecosystem is degraded when the ecosystem's ability to absorb a stress has been exceeded. A stress on an aquatic ecosystem can be a result of physical, chemical or biological alterations of the environment. Physical alterations include changes in water temperature, water flow and light availability. Chemical alterations include changes in the loading rates of biostimulatory nutrients, oxygen consuming materials, and toxins. Biological alterations include over-harvesting of commercial species and the introduction of exotic species. Human populations can impose excessive stresses on aquatic ecosystems. There are many examples of excessive stresses with negative consequences. Consider three. The environmental history of the Great Lakes of North America illustrates this problem, particularly how multiple stresses, such as water pollution, over-harvesting and invasive species can combine. The Norfolk Broadlands in England illustrate similar decline with pollution and invasive species. Lake Pontchartrain along the Gulf of Mexico illustrates the negative effects of different stresses including levee construction, logging of swamps, invasive species and salt water intrusion.

Abiotic Characteristics

An ecosystem is composed of biotic communities that are structured by biological interactions and abiotic environmental factors. Some of the important abiotic environmental factors of aquatic ecosystems include substrate type, water depth, nutrient levels, temperature, salinity, and flow. It is often difficult to determine the relative importance of these factors without rather large experiments. There may be complicated feedback loops. For example, sediment may determine the presence of aquatic plants, but aquatic plants may also trap sediment, and add to the sediment through peat.

The amount of dissolved oxygen in a water body is frequently the key substance in determining the extent and kinds of organic life in the water body. Fish need dissolved oxygen to survive, although their tolerance to low oxygen varies among species; in extreme cases of low oxygen some fish even resort to air gulping. Plants often have to produce aerenchyma, while the shape and size of leaves may also be altered. Conversely, oxygen is fatal to many kinds of anaerobic bacteria.

Nutrient levels are important in controlling the abundance of many species of algae. The relative abundance of nitrogen and phosphorus can in effect determine which species of algae come to dominate. Algae are a very important source of food for aquatic life, but at the same time, if they become over-abundant, they can cause declines in fish when they decay. Similar over-abundance of algae in coastal environments such as the Gulf of Mexico produces, upon decay, a hypoxic region of water known as a dead zone.

The salinity of the water body is also a determining factor in the kinds of species found in the water body. Organisms in marine ecosystems tolerate salinity, while many freshwater organisms are intolerant of salt. The degree of salinity in an estuary or delta is an important control upon the type of wetland (fresh, intermediate, or brackish), and the associated animal species. Dams built upstream may reduce spring flooding, and reduce sediment accretion, and may therefore lead to saltwater intrusion in coastal wetlands.

Freshwater used for irrigation purposes often absorbs levels of salt that are harmful to freshwater organisms.

Biotic Characteristics

The biotic characteristics are mainly determined by the organisms that occur. For example, wetland plants may produce dense canopies that cover large areas of sediment—or snails or geese may graze the vegetation leaving large mud flats. Aquatic environments have relatively low oxygen levels, forcing adaptation by the organisms found there. For example, many wetland plants must produce aerenchyma to carry oxygen to roots. Other biotic characteristics are more subtle and difficult to measure, such as the relative importance of competition, mutualism or predation. There are a growing number of cases where predation by coastal herbivores including snails, geese and mammals appears to be a dominant biotic factor.

Autotrophic Organisms

Autotrophic organisms are producers that generate organic compounds from inorganic material. Algae use solar energy to generate biomass from carbon dioxide and are possibly the most important autotrophic organisms in aquatic environments. Of course, the more shallow the water, the greater the biomass contribution from rooted and floating vascular plants. These two sources combine to produce the extraordinary production of estuaries and wetlands, as this autotrophic biomass is converted into fish, birds, amphibians and other aquatic species.

Chemosynthetic bacteria are found in benthic marine ecosystems. These organisms are able to feed on hydrogen sulfide in water that comes from volcanic vents. Great concentrations of animals that feed on these bacteria are found around volcanic vents. For example, there are giant tube worms (*Riftia pachyptila*) 1.5 m in length and clams (*Calyptogena magnifica*) 30 cm long.

Heterotrophic Organisms

Heterotrophic organisms consume autotrophic organisms and use the organic compounds in their bodies as energy sources and as raw materials to create their own biomass. Euryhaline organisms are salt tolerant and can survive in marine ecosystems, while stenohaline or salt intolerant species can only live in freshwater environments.

Freshwater Ecosystem

Freshwater ecosystems are a subset of Earth's aquatic ecosystems. They include lakes and ponds, rivers, streams, springs, and wetlands. They can be contrasted with marine ecosystems, which have a larger salt content. Freshwater habitats can be classified by different factors, including temperature, light penetration, and vegetation.

Freshwater ecosystems can be divided into lentic ecosystems (still water) and lotic ecosystems (flowing water).

Limnology (and its branch freshwater biology) is a study about freshwater ecosystems. It is a part of hydrobiology.

Original attempts to understand and monitor freshwater ecosystems were spurred on by threats to human health (ex. Cholera outbreaks due to sewage contamination). Early monitoring focussed on chemical indicators, then bacteria, and finally algae, fungi and protozoa. A new type of monitoring involves differing groups of organisms (macroinvertebrates, macrophytes and fish) and the stream conditions associated with them.

Current biomonitoring techniques focus mainly on community structure or biochemical oxygen demand. Responses are measured by behavioural changes, altered rates of growth, reproduction or mortality. Macroinvertebrates are most often used in these models because of well known taxonomy, ease of collection, sensitivity to a range of stressors, and their overall value to the ecosystem. Most of these measurements are difficult to extrapolate on a large scale, however.

The use of reference sites is common when assessing what a healthy freshwater ecosystem should "look like". Reference sites are easier to reconstruct in standing water than moving water. Preserved indicators such as diatom valves, macrophyte pollen, insect chitin and fish scales can be used to establish a reference ecosystem representative of a time before large scale human disturbance.

Common chemical stresses on freshwater ecosystem health include acidification, eutrophication and copper and pesticide contamination.

Extinction of Freshwater Fauna

Over 123 freshwater fauna species have gone extinct in North America since 1900. Of North American freshwater species, an estimated 48.5% of mussels, 22.8% of gastropods, 32.7% of crayfishes, 25.9% of amphibians, and 21.3% of fishes are either endangered or threatened. Extinction rates of many species may increase severely into the next century because of invasive species, loss of keystone species, and species which are already functionally extinct. Projected extinction rates for freshwater animals are around five times greater than for land animals, and are comparable to the rates for rainforest communities. Recent extinction trends can be attributed largely to sedimentation, stream fragmentation, chemical and organic pollutants, dams, and invasive species.

Lake Ecosystem

A lake ecosystem includes biotic (living) plants, animals and micro-organisms, as well as abiotic (nonliving) physical and chemical interactions.

Lake ecosystems are a prime example of lentic ecosystems. *Lentic* refers to stationary or relatively still water, from the Latin *lentus*, which means sluggish. Lentic waters range from ponds to lakes to wetlands. Lentic ecosystems can be compared with lotic ecosystems, which involve flowing terrestrial waters such as rivers and streams. Together, these two fields form the more general study area of freshwater or aquatic ecology.

Lentic systems are diverse, ranging from a small, temporary rainwater pool a few inches deep to Lake Baikal, which has a maximum depth of 1740 m. The general distinction between pools/ponds and lakes is vague, but Brown states that ponds and pools have their entire bottom surfaces exposed to light, while lakes do not. In addition, some lakes become seasonally stratified (discussed in more detail below.) Ponds and pools have two regions: the pelagic open water zone, and the benthic zone, which comprises the bottom and shore regions. Since lakes have deep bottom regions not exposed to light, these systems have an additional zone, the profundal. These three areas can have very different abiotic conditions and, hence, host species that are specifically adapted to live there.

Important Abiotic Factors

Light

Light provides the solar energy required to drive the process of photosynthesis, the major energy source of lentic systems. The amount of light received depends upon a combination of several factors. Small ponds may experience shading by surrounding trees, while cloud cover may affect light availability in all systems, regardless of size. Seasonal and diurnal considerations also play a role in light availability because the shallower the angle at which light strikes water, the more light is lost by reflection. This is known as Beer's law. Once light has penetrated the surface, it may also be scattered by particles suspended in the water column. This scattering decreases the total amount of light as depth increases. Lakes are divided into photic and aphotic regions, the prior receiving sunlight and latter being below the depths of light penetration, making it void of photosynthetic capacity. In relation to lake zonation, the pelagic and benthic zones are considered to lie within the photic region, while the profundal zone is in the aphotic region.

Temperature

Temperature is an important abiotic factor in lentic ecosystems because most of the biota are poikilothermic, where internal body temperatures are defined by the surrounding system. Water can be heated or cooled through radiation at the surface and conduction to or from the air and surrounding substrate. Shallow ponds often have a continuous temperature gradient from warmer waters at the surface to cooler waters at the bottom. In addition, temperature fluctuations can be very great in these systems, both diurnally and seasonally.

Temperature regimes are very different in large lakes. In temperate regions, for example, as air temperatures increase, the icy layer formed on the surface of the lake breaks up, leaving the water at approximately 4 °C. This is the temperature at which water has the highest density. As the season progresses, the warmer air temperatures heat the surface waters, making them less dense. The deeper waters remain cool and dense due to reduced light penetration. As the summer begins, two distinct layers become established, with such a large temperature difference between them that they remain stratified. The lowest zone in the lake is the coldest and is called the hypolimnion. The upper warm zone is called the epilimnion. Between these zones is a band of rapid temperature change called the thermocline. During the colder fall season, heat is lost at the surface and the epilimnion cools. When the temperatures of the two zones are close enough, the waters begin to mix again to create a uniform temperature, an event termed lake turnover. In the winter, inverse stratification occurs as water near the surface cools freezes, while warmer, but denser water remains near the bottom. A thermocline is established, and the cycle repeats.

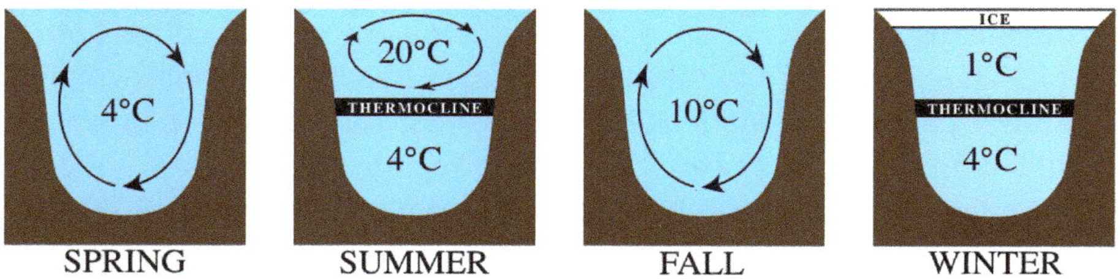

Seasonal stratification in temperate lakes

Wind

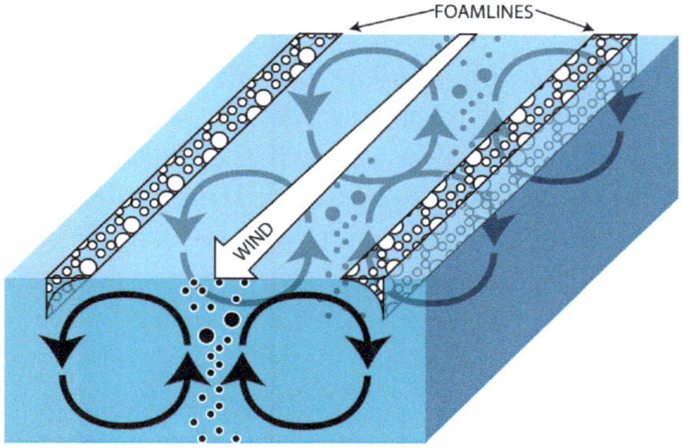

Illustration of Langmuir rotations; open circles=positively buoyant particles, closed circles=negatively buoyant particles

In exposed systems, wind can create turbulent, spiral-formed surface currents called Langmuir circulations. Exactly how these currents become established is still not well understood, but it is evident that it involves some interaction between horizontal surface currents and surface gravity waves. The visible result of these rotations, which can be seen in any lake, are the surface foamlines that run parallel to the wind direction. Positively buoyant particles and small organisms concentrate in the foamline at the surface and negatively buoyant objects are found in the upwelling current between the two rotations. Objects with neutral buoyancy tend to be evenly distributed in the water column. This turbulence circulates nutrients in the water column, making it crucial for many pelagic species, however its effect on benthic and profundal organisms is minimal to non-existent, respectively. The degree of nutrient circulation is system specific, as it depends upon such factors as wind strength and duration, as well as lake or pool depth and productivity.

Chemistry

Oxygen is essential for organismal respiration. The amount of oxygen present in standing waters depends upon: 1) the area of transparent water exposed to the air, 2) the circulation of water within the system and 3) the amount of oxygen generated and used by organisms present. In shallow, plant-rich pools there may be great fluctuations of oxygen, with extremely high concentrations occurring during the day due to photosynthesis and very low values at night when respiration is the dominant process of primary producers. Thermal stratification in larger systems can also affect the amount of oxygen present in different zones. The epilimnion is oxygen rich because it circulates quickly, gaining oxygen via contact with the air. The hypolimnion, however, circulates very slowly and has no atmospheric contact. Additionally, fewer green plants exist in the hypolimnion, so there is less oxygen released from photosynthesis. In spring and fall when the epilimnion and hypolimnion mix, oxygen becomes more evenly distributed in the system. Low oxygen levels are characteristic of the profundal zone due to the accumulation of decaying vegetation and animal matter that "rains" down from the pelagic and benthic zones and the inability to support primary producers.

Phosphorus is important for all organisms because it is a component of DNA and RNA and is involved in cell metabolism as a component of ATP and ADP. Also, phosphorus is not found in large quantities in freshwater systems, limiting photosynthesis in primary producers, making it the main determinant of lentic system production. The phosphorus cycle is complex, but the model outlined below describes the basic pathways. Phosphorus mainly enters a pond or lake through runoff from the watershed or by atmospheric deposition. Upon entering the system, a reactive form of phosphorus is usually taken up by algae and macrophytes, which release a non-reactive phosphorus compound as a byproduct of photosynthesis. This phosphorus can drift downwards and become part of the benthic or profundal sediment, or it can be remineralized to the reactive form by microbes in the water column. Similarly, non-reactive phosphorus in the sediment can be remineralized into the reactive form. Sediments are generally richer in phosphorus than lake water, however, indicating that this nutrient may have a long residency time there before it is remineralized and re-introduced to the system.

Lentic System Biota

Bacteria

Bacteria are present in all regions of lentic waters. Free-living forms are associated with decomposing organic material, biofilm on the surfaces of rocks and plants, suspended in the water column, and in the sediments of the benthic and profundal zones. Other forms are also associated with the guts of lentic animals as parasites or in commensal relationships. Bacteria play an important role in system metabolism through nutrient recycling, which is discussed in the Trophic Relationships section.

Primary Producers

Nelumbo nucifera, an aquatic plant.

Algae, including both phytoplankton and periphyton are the principle photosynthesizers in ponds and lakes. Phytoplankton are found drifting in the water column of the pelagic zone. Many species have a higher density than water which should make them sink and end up in the benthos. To combat this, phytoplankton have developed density changing mechanisms, by forming vacuoles and gas vesicles or by changing their shapes to induce drag, slowing their descent. A very sophisticated adaptation utilized by a small number of species is a tail-like flagellum that can adjust vertical position and allow movement in any direction. Phytoplankton can also maintain their presence in the water column by being circulated in Langmuir rotations. Periphytic algae, on the other hand, are attached to a substrate. In lakes and ponds, they can cover all benthic surfaces. Both types of plankton are important as food sources and as oxygen providers.

Aquatic plants live in both the benthic and pelagic zones and can be grouped according to their manner of growth: 1) emergent = rooted in the substrate but with leaves and flowers extending into the air, 2) floating-leaved = rooted in the substrate but with floating leaves, 3) submersed = growing beneath the surface and 4) free-floating macrophytes = not rooted in the substrate and floating on the surface. These various forms of macrophytes generally occur in different areas of the benthic zone, with emergent vegetation nearest the shoreline, then floating-leaved macrophytes, followed by submersed vegetation. Free-floating macrophytes can occur anywhere on the system's surface.

Aquatic plants are more buoyant than their terrestrial counterparts because freshwater has a higher density than air. This makes structural rigidity unimportant in lakes and ponds (except in the aerial stems and leaves). Thus, the leaves and stems of most aquatic plants use less energy to construct and maintain woody tissue, investing that energy into fast growth instead. In order to contend with stresses induced by wind and waves, plants must be both flexible and tough. Light, water depth and substrate types are the most important factors controlling the distribution of submerged aquatic plants. Macrophytes are sources of food, oxygen, and habitat structure in the benthic zone, but cannot penetrate the depths of the euphotic zone and hence are not found there.

Invertebrates

Water striders are predatory insects which rely on surface tension to walk on top of water.
They live on the surface of ponds, marshes, and other quiet waters. They can move very quickly, up to 1.5 m/s.

Zooplankton are tiny animals suspended in the water column. Like phytoplankton, these species have developed mechanisms that keep them from sinking to deeper waters, including drag-inducing body forms and the active flicking of appendages such as antennae or spines. Remaining in the water column may have its advantages in terms of feeding, but this zone's lack of refugia leaves zooplankton vulnerable to predation. In response, some species, especially Daphnia sp., make daily vertical migrations in the water column by passively sinking to the darker lower depths during the day and actively moving towards the surface during the night. Also, because conditions in a lentic system can be quite variable across seasons, zooplankton have the ability to switch from laying regular eggs to resting eggs when there is a lack of food, temperatures fall below 2 °C, or if predator abundance is high. These resting eggs have a diapause, or dormancy period that should allow the zooplankton to encounter conditions that are more favorable to survival when they finally hatch. The invertebrates that inhabit the benthic zone are numerically dominated by small species and are species rich compared to the zooplankton of the open water. They include Crustaceans (e.g. crabs, crayfish, and shrimp), molluscs (e.g. clams and snails), and numerous types of insects. These organisms are mostly found in the areas of macrophyte growth, where the richest resources, highly oxygenated water, and warmest portion of the ecosystem are found. The structurally diverse macrophyte beds are important sites for the accumulation of organic matter, and provide an ideal area for colonization. The sediments and plants also offer a great deal of protection from predatory fishes.

Very few invertebrates are able to inhabit the cold, dark, and oxygen poor profundal zone. Those that can are often red in color due to the presence of large amounts of hemoglobin, which greatly increases the amount of oxygen carried to cells. Because the concentration of oxygen within this zone is low, most species construct tunnels or borrows in which they can hide and make the minimum movements necessary to circulate water through, drawing oxygen to them without expending much energy.

Fish and Other Vertebrates

Fish have a range of physiological tolerances that are dependent upon which species they belong to. They have different lethal temperatures, dissolved oxygen requirements, and spawning needs that are based on their activity levels and behaviors. Because fish are highly mobile, they are able to deal with unsuitable abiotic factors in one zone by simply moving to another. A detrital feeder in the profundal zone, for example, that finds the oxygen concentration has dropped too low may feed closer to the benthic zone. A fish might also alter its residence during different parts of its life history: hatching in a sediment nest, then moving to the weedy benthic zone to develop in a protected environment with food resources, and finally into the pelagic zone as an adult.

Other vertebrate taxa inhabit lentic systems as well. These include amphibians (e.g. salamanders and frogs), reptiles (e.g. snakes, turtles, and alligators), and a large number of waterfowl species. Most of these vertebrates spend part of their time in terrestrial habitats and thus are not directly affected by abiotic factors in the lake or pond. Many fish species are important as consumers and as prey species to the larger vertebrates mentioned above.

Trophic Relationships

Primary Producers

Lentic systems gain most of their energy from photosynthesis performed by aquatic plants and algae. This autochthonous process involves the combination of carbon dioxide, water, and solar energy to produce carbohydrates and dissolved oxygen. Within a lake or pond, the potential rate of photosynthesis generally decreases with depth due to light attenuation. Photosynthesis, however, is often low at the top few millimeters of the surface, likely due to inhibition by ultraviolet light. The exact depth and photosynthetic rate measurements of this curve are system specific and depend upon: 1) the total biomass of photosynthesizing cells, 2) the amount of light attenuating materials and 3) the abundance and frequency range of light absorbing pigments (i.e. chlorophylls) inside of photosynthesizing cells. The energy created by these primary producers is important for the community because it is transferred to higher trophic levels via consumption.

Bacteria

The vast majority of bacteria in lakes and ponds obtain their energy by decomposing vegetation and animal matter. In the pelagic zone, dead fish and the occasional allochthonous input of litterfall are examples of coarse particulate organic matter (CPOM>1 mm). Bacteria degrade these into fine particulate organic matter (FPOM<1 mm) and then further into usable nutrients. Small organisms such as plankton are also characterized as FPOM. Very low concentrations of nutrients are released during decomposition because the bacteria are utilizing them to build their own biomass. Bacteria, however, are consumed by protozoa, which are in turn consumed by zooplankton, and then further up the trophic levels. Nutrients, including those that contain carbon and phosphorus, are reintroduced into the water column at any number of points along this food chain via excretion or organism death, making them available again for bacteria. This regeneration cycle is known as the microbial loop and is a key component of lentic food webs.

The decomposition of organic materials can continue in the benthic and profundal zones if the matter falls through the water column before being completely digested by the pelagic bacteria. Bacteria are found in the greatest abundance here in sediments, where they are typically 2-1000 times more prevalent than in the water column.

Benthic Invertebrates

Benthic invertebrates, due to their high level of species richness, have many methods of prey capture. Filter feeders create currents via siphons or beating cilia, to pull water and its nutritional contents, towards themselves for straining. Grazers use scraping, rasping, and shredding adaptations to feed on periphytic algae and macrophytes. Members of the collector guild browse the sediments, picking out specific particles with raptorial appendages. Deposit feeding invertebrates indiscriminately consume sediment, digesting any organic material it contains. Finally, some invertebrates belong to the predator guild, capturing and consuming living animals. The profundal zone is home to a unique group of filter feeders that use small body movements to draw a current through burrows that they have created in the sediment. This mode of feeding requires the least amount of motion, allowing these species to conserve energy. A small number of invertebrate taxa are predators in the profundal zone. These species are likely from other regions and only come to these depths to feed. The vast majority of invertebrates in this zone are deposit feeders, getting their energy from the surrounding sediments.

Fish

Fish size, mobility, and sensory capabilities allow them to exploit a broad prey base, covering multiple zonation regions. Like invertebrates, fish feeding habits can be categorized into guilds. In the pelagic zone, herbivores graze on periphyton and macrophytes or pick phytoplankton out of the water column. Carnivores include fishes that feed on zooplankton in the water column (zooplanktivores), insects at the water's surface, on benthic structures, or in the sediment (insectivores), and those that feed on other fish (piscivores). Fish that consume detritus and gain energy by processing its organic material are called detritivores. Omnivores ingest a wide variety of prey, encompassing floral, faunal, and detrital material. Finally, members of the parasitic guild acquire nutrition from a host species, usually another fish or large vertebrate. Fish taxa are flexible in their feeding roles, varying their diets with environmental conditions and prey availability. Many species also undergo a diet shift as they develop. Therefore, it is likely that any single fish occupies multiple feeding guilds within its lifetime.

Lentic Food Webs

The lentic biota are linked in complex web of trophic relationships. These organisms can be considered to loosely be associated with specific trophic groups (e.g. primary producers, herbivores, primary carnivores, secondary carnivores, etc.). Scientists have developed several theories in order to understand the mechanisms that control the abundance and diversity within these groups. Very generally, top-down processes dictate that the abundance of prey taxa is dependent upon the actions of consumers from higher trophic levels. Typically, these processes operate only between two trophic levels, with no effect on the others. In some cases, however, aquatic systems experience a trophic cascade; for example, this might occur if primary producers experience less grazing by herbivores because these herbivores are suppressed by carnivores. Bottom-up processes are functioning when the abundance or diversity of members of higher trophic levels is dependent upon the availability or quality of resources from lower levels. Finally, a combined regulating theory, bottom-up:top-down, combines the predicted influences of consumers and resource availability. It predicts that trophic levels close to the lowest trophic levels will be most influenced by bottom-up forces, while top-down effects should be strongest at top levels.

Community Patterns and Diversity

Local Species Richness

The biodiversity of a lentic system increases with the surface area of the lake or pond. This is attributable to the higher likelihood of partly terrestrial species of finding a larger system. Also, because larger systems typically have larger populations, the chance of extinction is decreased. Additional factors, including temperature regime, pH, nutrient availability, habitat complexity, speciation rates, competition, and predation, have been linked to the number of species present within systems.

Succession Patterns in Plankton Communities – the PEG Model

Phytoplankton and zooplankton communities in lake systems undergo seasonal succession in relation to nutrient availability, predation, and competition. Sommer *et al.* described these patterns as part of the Plankton Ecology Group (PEG) model, with 24 statements constructed from the analysis of numerous systems. The following includes a subset of these statements, as explained by Brönmark and Hansson illustrating succession through a single seasonal cycle:

Winter

1. Increased nutrient and light availability result in rapid phytoplankton growth towards the end of winter. The dominant species, such as diatoms, are small and have quick growth capabilities. 2. These plankton are consumed by zooplankton, which become the dominant plankton taxa.

Spring

3. A clear water phase occurs, as phytoplankton populations become depleted due to increased predation by growing numbers of zooplankton.

Summer

4. Zooplankton abundance declines as a result of decreased phytoplankton prey and increased predation by juvenile fishes.

5. With increased nutrient availability and decreased predation from zooplankton, a diverse phytoplankton community develops.

6. As the summer continues, nutrients become depleted in a predictable order: phosphorus, silica, and then nitrogen. The abundance of various phytoplankton species varies in relation to their biological need for these nutrients.

7. Small-sized zooplankton become the dominant type of zooplankton because they are less vulnerable to fish predation.

Fall

8. Predation by fishes is reduced due to lower temperatures and zooplankton of all sizes increase in number.

Winter

9. Cold temperatures and decreased light availability result in lower rates of primary production and decreased phytoplankton populations. 10. Reproduction in zooplankton decreases due to lower temperatures and less prey.

The PEG model presents an idealized version of this succession pattern, while natural systems are known for their variation.

Latitudinal Patterns

There is a well-documented global pattern that correlates decreasing plant and animal diversity with increasing latitude, that is to say, there are fewer species as one moves towards the poles. The cause of this pattern is one of the greatest puzzles for ecologists today. Theories for its explanation include energy availability, climatic variability, disturbance, competition, etc. Despite this global diversity gradient, this pattern can be weak for freshwater systems compared to global marine and terrestrial systems. This may be related to size, as Hillebrand and Azovsky found that smaller organisms (protozoa and plankton) did not follow the expected trend strongly, while larger species (vertebrates) did. They attributed this to better dispersal ability by smaller organisms, which may result in high distributions globally.

Natural Lake Lifecycles

Lake Creation

Lakes can be formed in a variety of ways, but the most common are discussed briefly below. The oldest and largest systems are the result of tectonic activities. The rift lakes in Africa, for example are the result of seismic activity along the site of separation of two tectonic plates. Ice-formed lakes are created when glaciers recede, leaving behind abnormalities in the landscape shape that are then filled with water. Finally, oxbow lakes are fluvial in origin, resulting when a meandering river bend is pinched off from the main channel.

Natural Extinction

All lakes and ponds receive sediment inputs. Since these systems are not really expanding, it is logical to assume that they will become increasingly shallower in depth, eventually becoming wetlands or terrestrial vegetation. The length of this process should depend upon a combination of depth and sedimentation rate. Moss gives the example of Lake Tanganyika, which reaches a depth of 1500 m and has a sedimentation rate of 0.5 mm/yr. Assuming that sedimentation is not influenced by anthropogenic factors, this system should go extinct in approximately 3 million years. Shallow lentic systems might also fill in as swamps encroach inward from the edges. These processes operate on a much shorter timescale, taking hundreds to thousands of years to complete the extinction process.

Human Impacts

Acidification

Sulfur dioxide and nitrogen oxides are naturally released from volcanoes, organic compounds in the soil, wetlands, and marine systems, but the majority of these compounds come from the combustion of coal, oil, gasoline, and the smelting of ores containing sulfur. These substances dissolve in atmospheric moisture and enter lentic systems as acid rain. Lakes and ponds that contain bedrock that is rich in carbonates have a natural buffer, resulting in no alteration of pH. Systems without this bedrock, however, are very sensitive to acid inputs because they have a low neutralizing capacity, resulting in pH declines even with only small inputs of acid. At a pH of 5–6 algal species diversity and biomass decrease considerably, leading to an increase in water transparency – a characteristic feature of acidified lakes. As the pH continues lower, all fauna becomes less diverse. The most significant feature is the disruption of fish reproduction. Thus, the population is eventually composed of few, old individuals that eventually die and leave the systems without fishes. Acid rain has been especially harmful to lakes in Scandinavia, western Scotland, west Wales and the north eastern United States.

Eutrophication

Eutrophic systems contain a high concentration of phosphorus (~30 µg/L), nitrogen (~1500 µg/L), or both. Phosphorus enters lentic waters from sewage treatment effluents, discharge from raw sewage, or from runoff of farmland. Nitrogen mostly comes from agricultural fertilizers from runoff or leaching and subsequent groundwater flow. This increase in nutrients required for primary producers results in a massive increase of phytoplankton growth, termed a plankton bloom. This bloom decreases water transparency, leading to the loss of submerged plants. The resultant reduction in habitat structure has negative impacts on the species' that utilize it for spawning, maturation and general survival. Additionally, the large number of short-lived phytoplankton result in a massive amount of dead biomass settling into the sediment. Bacteria need large amounts of oxygen to decompose this material, reducing the oxygen concentration of the water. This is especially pronounced in stratified lakes when the thermocline prevents oxygen rich water from the surface to mix with lower levels. Low or anoxic conditions preclude the existence of many taxa that are not physiologically tolerant of these conditions.

Invasive Species

Invasive species have been introduced to lentic systems through both purposeful events (e.g. stocking game and food species) as well as unintentional events (e.g. in ballast water). These organisms can affect natives via competition for prey or habitat, predation, habitat alteration, hybridization, or the introduction of harmful diseases and parasites. With regard to native species, invaders may cause changes in size and age structure, distribution, density, population growth, and may even drive populations to extinction. Examples of prominent invaders of lentic systems include the zebra mussel and sea lamprey in the Great Lakes.

Marine Ecosystem

Marine ecosystems are among the largest of Earth's aquatic ecosystems. Examples include salt marshes, intertidal zones, estuaries, lagoons, mangroves, coral reefs, the deep sea, and the sea floor. They can be contrasted with freshwater ecosystems, which have a lower salt content. Marine waters cover two-thirds of the surface of the Earth. Such places are considered ecosystems because the plant life supports the animal life and vice versa.

Coral reefs form complex marine ecosystems with tremendous biodiversity.

Here, we can see different types of starfish, coral reefs and fishes in the Great Barrier Reef.

Marine ecosystems are essential for the overall health of both marine and terrestrial environments. According to the World Resource Center, coastal habitats account for about one-third of marine biological productivity. Estuarine ecosystems, such as salt marshes, seagrass meadows and mangrove forests, are among the most productive ecosystems on the planet. Coral reefs provide food and shelter to the highest levels of marine diversity in the world.

Marine ecosystems usually have a large biodiversity and are therefore thought to have a good resistance against invasive species. However, exceptions have been observed, and the mechanisms responsible in determining the success of an invasion are not yet clear.

Marine Habitats

The marine environment supplies many kinds of habitats that support marine life. Marine life depends in some way on the saltwater that is in the sea (the term *marine* comes from the Latin *mare*, meaning sea or ocean). A habitat is an ecological or environmental area inhabited by one or more living species.

Marine habitats can be divided into coastal and open ocean habitats. Coastal habitats are found in the area that extends from as far as the tide comes in on the shoreline out to the edge of the continental shelf. Most marine life is found in coastal habitats, even though the shelf area occupies only seven percent of the total ocean area. Open ocean habitats are found in the deep ocean beyond the edge of the continental shelf.

Alternatively, marine habitats can be divided into pelagic and demersal zones. Pelagic habitats are found near the surface or in the open water column, away from the bottom of the ocean. Demersal habitats are near or on the bottom of the ocean. An organism living in a pelagic habitat is said to be a pelagic organism, as in pelagic fish. Similarly, an organism living in a demersal habitat is said to be a demersal organism, as in demersal fish. Pelagic habitats are intrinsically shifting and ephemeral, depending on what ocean currents are doing.

Marine habitats can be modified by their inhabitants. Some marine organisms, like corals, kelp, mangroves and seagrasses, are ecosystem engineers which reshape the marine environment to the point where they create further habitat for other organisms.

Overview

Only 29 percent of the world surface is land. The rest is ocean, home to the marine habitats. The oceans are nearly four kilometres deep on average and are fringed with coastlines that run for nearly 380,000 kilometres.

In contrast to terrestrial habitats, marine habitats are shifting and ephemeral. Swimming organisms find areas by the edge of a continental shelf a good habitat, but only while upwellings bring nutrient rich water to the surface. Shellfish find habitat on sandy beaches, but storms, tides and currents mean their habitat continually reinvents itself.

The presence of seawater is common to all marine habitats. Beyond that many other things determine whether a marine area makes a good habitat and the type of habitat it makes. For example:

- Temperature – is affected by geographical latitude, ocean currents, weather, the discharge of rivers, and by the presence of hydrothermal vents or cold seeps.
- Sunlight – photosynthetic processes depend on how deep and turbid the water is.
- Nutrients – are transported by ocean currents to different marine habitats from land run-off, or by upwellings from the deep sea, or they sink though the sea as marine snow.
- Salinity – varies, particularly in estuaries or near river deltas, or by hydrothermal vents.
- Dissolved gases – oxygen levels in particular, can be increased by wave actions and decreased during algal blooms.
- Acidity – this is partly to do with dissolved gases above, since the acidity of the ocean is largely controlled by how much carbon dioxide is in the water.
- Turbulence – ocean waves, fast currents and the agitation of water affect the nature of habitats.
- Cover – the availability of cover such as the adjacency of the sea bottom, or the presence of floating objects.
- The occupying organisms themselves – since organisms modify their habitats by the act of occupying them, and some, like corals, kelp, mangroves and seagrasses, create further habitats for other organisms.

Ocean	Area million km²	%	Volume million cu km	%	Mean depth km	Max depth km	Coastline km
Pacific Ocean	155.6	46.4	679.6	49.6	4.37	10.924	135,663
Atlantic Ocean	76.8	22.9	313.4	22.5	4.08	8.605	111,866
Indian Ocean	68.6	20.4	269.3	19.6	3.93	7.258	66,526
Southern Ocean	20.3	6.1	91.5	6.7	4.51	7.235	17,968
Arctic Ocean	14.1	4.2	17.0	1.2	1.21	4.665	45,389
Overall	**335.3**		**1370.8**		**4.09**	**10.924**	**377,412**

The ocean occupies 71 percent of the world surface, averaging nearly four kilometres in depth. There are five major oceans, of which the Pacific Ocean is nearly as large as the rest put together. Coastlines fringe the land for nearly 380,000 kilometres.

Land runoff, pouring into the sea, can contain nutrients

Marine habitats can be broadly divided into pelagic and demersal habitats. Pelagic habitats are the habitats of the open water column, away from the bottom of the ocean. Demersal habitats are the habitats that are near or on the bottom of the ocean. An organism living in a pelagic habitat is said to be a pelagic organism, as in pelagic fish. Similarly, an organism living in a demersal habitat is said to be a demersal organism, as in demersal fish. Pelagic habitats are intrinsically ephemeral, depending on what ocean currents are doing.

The land-based ecosystem depends on topsoil and fresh water, while the marine ecosystem depends on dissolved nutrients washed down from the land.

Ocean deoxygenation poses a threat to marine habitats, due to the growth of low oxygen zones.

Ocean Currents

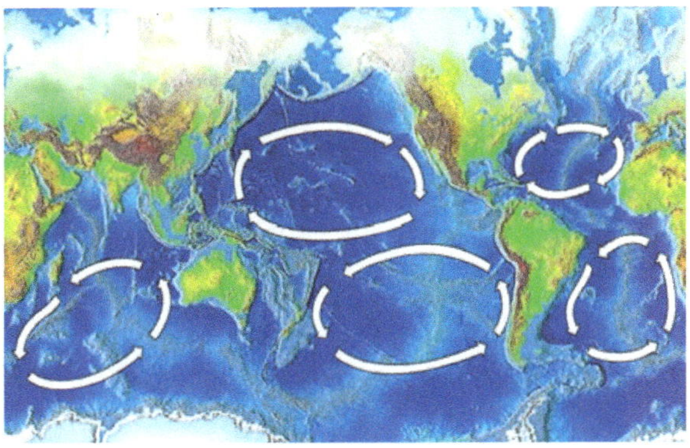

Ocean gyres rotate clockwise in the north and counterclockwise in the south

In marine systems, ocean currents have a key role determining which areas are effective as habitats, since ocean currents transport the basic nutrients needed to support marine life. Plankton are the life forms that inhabit the ocean that are so small (less than 2 mm) that they cannot effectively propel themselves through the water, but must drift instead with the currents. If the current carries the right nutrients, and if it also flows at a suitably shallow depth where there is plenty of sunlight, then such a current itself can become a suitable habitat for photosynthesizing tiny algae called phytoplankton. These tiny plants are the primary producers in the ocean, at the start of the food chain. In turn, as the population of drifting phytoplankton grows, the water becomes a suitable habitat for zooplankton, which feed on the phytoplankton. While phytoplankton are tiny drifting plants, zooplankton are tiny drifting animals, such as the larvae of fish and marine invertebrates. If sufficient zooplankton establish themselves, the current becomes a candidate habitat for the forage fish that feed on them. And then if sufficient forage fish move to the area, it becomes a candidate habitat for larger predatory fish and other marine animals that feed on the forage fish. In this dynamic way, the current itself can, over time, become a moving habitat for multiple types of marine life.

This algae bloom occupies sunlit epipelagic waters off the southern coast of England. The algae are maybe feeding on nutrients from land runoff or upwellings at the edge of the continental shelf

Ocean currents can be generated by differences in the density of the water. How dense water is depends on how saline or warm it is. If water contains differences in salt content or temperature, then the different densities will initiate a current. Water that is saltier or cooler will be denser, and will sink in relation to the surrounding water. Conversely, warmer and less salty water will float to the surface. Atmospheric winds and pressure differences also produces surface currents, waves and seiches. Ocean currents are also generated by the gravitational pull of the sun and moon (tides), and seismic activity (tsunami).

The rotation of the Earth affects the direction ocean currents take, and explains which way the large circular ocean gyres rotate in the image above left. Suppose a current at the equator is heading north. The Earth rotates eastward, so the water possesses that rotational momentum. But the further the water moves north, the slower the earth moves eastward. If the current could get to the North Pole, the earth wouldn't be moving eastward at all. To conserve its rotational momentum, the further the current travels north the faster it must move eastward. So the effect is that the current curves to the right. This is the Coriolis effect. It is weakest at the equator and strongest at the poles. The effect is opposite south of the equator, where currents curve left.

Marine Topography

Marine (or seabed or ocean) topography refers to the shape the land has when it interfaces with the ocean. These shapes are obvious along coastlines, but they occur also in significant ways underwater. The effectiveness of marine habitats is partially defined by these shapes, including the way they interact with and shape ocean currents, and the way sunlight diminishes when these landforms occupy increasing depths. Tidal networks depend on the balance between sedimentary processes and hydrodynamics however, anthropogenic influences can impact the natural system more than any physical driver.

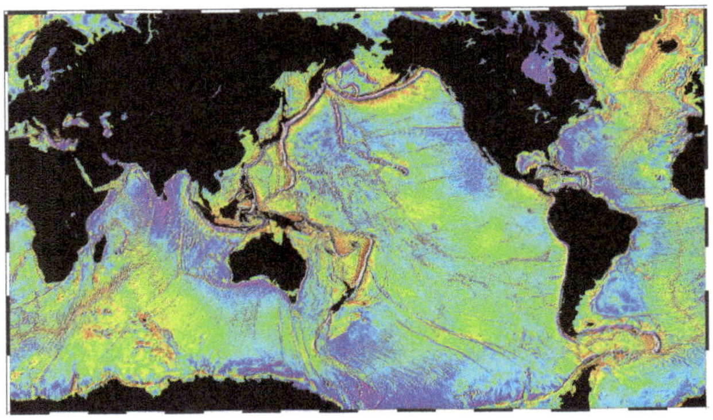

Map of underwater topography (1995 NOAA)

Marine topographies include coastal and oceanic landforms ranging from coastal estuaries and shorelines to continental shelves and coral reefs. Further out in the open ocean, they include underwater and deep sea features such as ocean rises and seamounts. The submerged surface has mountainous features, including a globe-spanning mid-ocean ridge system, as well as undersea volcanoes, oceanic trenches, submarine canyons, oceanic plateaus and abyssal plains.

The mass of the oceans is approximately 1.35×10^{18} metric tons, or about 1/4400 of the total mass of the Earth. The oceans cover an area of 3.618×10^{8} km² with a mean depth of 3,682 m, resulting in an estimated volume of 1.332×10^{9} km³.

Biomass

One measure of the relative importance of different marine habitats is the rate at which they produce biomass.

Producer	Biomass productivity (gC/m²/yr)	Total area (million km²)	Total production (billion tonnes C/yr)	Comment
swamps and marshes	2,500			Includes freshwater
coral reefs	2,000	0.28	0.56	
algal beds	2,000			
river estuaries	1,800			
open ocean	125	311	39	

Coastal

Marine coasts are dynamic environments which constantly change, like the ocean which partially shape them. The Earth's natural processes, including weather and sea level change, result in the erosion, accretion and resculpturing of coasts as well as the flooding and creation of continental shelves and drowned river valleys.

The main agents responsible for deposition and erosion along coastlines are waves, tides and currents. The formation of coasts also depends on the nature of the rocks they are made of – the harder the rocks the less likely they are to erode, so variations in rock hardness result in coastlines with different shapes.

Coastlines can be volatile habitats

Tides often determine the range over which sediment is deposited or eroded. Areas with high tidal ranges allow waves to reach farther up the shore, and areas with lower tidal ranges produce deposition at a smaller elevation interval. The tidal range is influenced by the size and shape of the coastline. Tides do not typically cause erosion by themselves; however, tidal bores can erode as the waves surge up river estuaries from the ocean.

Waves erode coastline as they break on shore releasing their energy; the larger the wave the more energy it releases and the more sediment it moves. Sediment deposited by waves comes from eroded cliff faces and is moved along the coastline by the waves. Sediment deposited by rivers is the dominant influence on the amount of sediment located on a coastline.

The sedimentologist Francis Shepard classified coasts as *primary* or *secondary*.

- Primary coasts are shaped by non-marine processes, by changes in the land form. If a coast is in much the same condition as it was when sea level was stabilised after the last ice age, it is called a primary coast. "Primary coasts are created by erosion (the wearing away of soil or rock), deposition (the buildup of sediment or sand) or tectonic activity (changes in the structure of the rock and soil because of earthquakes). Many of these coastlines were formed as the sea level rose during the last 18,000 years, submerging river and glacial valleys to form bays and fjords." An example of a primary coast is a river delta, which forms when a river deposits soil and other material as it enters the sea.

- Secondary coasts are produced by marine processes, such as the action of the sea or by creatures that live in it. Secondary coastlines include sea cliffs, barrier islands, mud flats, coral reefs, mangrove swamps and salt marshes.

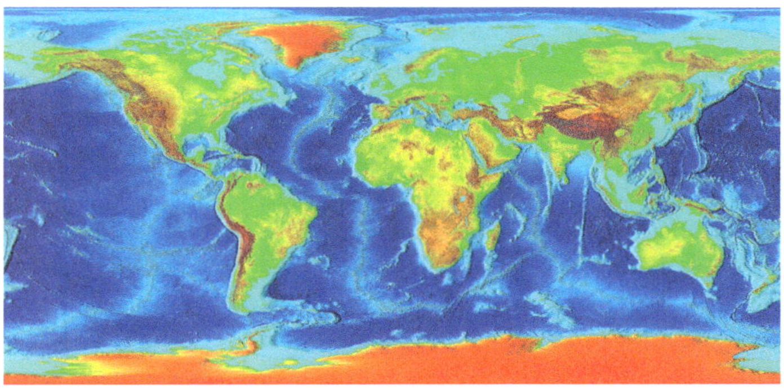

The global continental shelf, highlighted in cyan,
defines the extent of coastal habitats, and occupies 5% of the total world area.

Continental coastlines usually have a continental shelf, a shelf of relatively shallow water, less than 200 metres deep, which extends 68 km on average beyond the coast. Worldwide, continental shelves occupy a total area of about 24 million km^2 (9 million sq mi), 8% of the ocean's total area and nearly 5% of the world's total area. Since the continental shelf is usually less than 200 metres deep, it follows that coastal habitats are generally photic, situated in the sunlit epipelagic zone. This means the conditions for photosynthetic processes so important for primary production, are available to coastal marine habitats. Because land is nearby, there are large discharges of nutrient rich land runoff into coastal waters. Further, periodic upwellings from the deep ocean can provide cool and nutrient rich currents along the edge of the continental shelf.

As a result, coastal marine life is the most abundant in the world. It is found in tidal pools, fjords and estuaries, near sandy shores and rocky coastlines, around coral reefs and on or above the continental shelf. Coastal fish include small forage fish as well as the larger predator fish that feed on them. Forage fish thrive in inshore waters where high productivity results from upwelling and shoreline run off of nutrients. Some are partial residents that spawn in streams, estuaries and bays, but most complete their life cycle in the zone. There can also be a mutualism between species that occupy adjacent marine habitats. For example, fringing reefs just below low tide level have a mutually beneficial relationship with mangrove forests at high tide level and sea grass meadows in between: the reefs protect the mangroves and seagrass from strong currents and waves that would damage them or erode the sediments in which they are rooted, while the mangroves and seagrass protect the coral from large influxes of silt, fresh water and pollutants. This additional level of variety in the environment is beneficial to many types of coral reef animals, which for example may feed in the sea grass and use the reefs for protection or breeding.

Coastal habitats are the most visible marine habitats, but they are not the only important marine habitats. Coastlines run for 380,000 kilometres, and the total volume of the ocean is 1,370 million cu km. This means that for each metre of coast, there is 3.6 cu km of ocean space available somewhere for marine habitats.

Waves and currents shape the intertidal shoreline,
eroding the softer rocks and transporting and grading loose particles into shingles, sand or mud

Intertidal

Intertidal zones, those areas close to shore, are constantly being exposed and covered by the ocean's tides. A huge array of life lives within this zone.

Shore habitats range from the upper intertidal zones to the area where land vegetation takes prominence. It can be underwater anywhere from daily to very infrequently. Many species here are scavengers, living off of sea life that is washed up on the shore. Many land animals also make much use of the shore and intertidal habitats. A subgroup of organisms in this habitat bores and grinds exposed rock through the process of bioerosion.

Sandy Shores

Sandy shores provide shifting homes to many species

Sandy shores, also called beaches, are coastal shorelines where sand accumulates. Waves and currents shift the sand, continually building and eroding the shoreline. Longshore currents flow parallel to the beaches, making waves break obliquely on the sand. These currents transport large amounts of sand along coasts, forming spits, barrier islands and tombolos. Longshore currents also commonly create offshore bars, which give beaches some stability by reducing erosion.

Sandy shores are full of life, The grains of sand host diatoms, bacteria and other microscopic creatures. Some fish and turtles return to certain beaches and spawn eggs in the sand. Birds habitat beaches, like gulls, loons, sandpipers, terns and pelicans. Aquatic mammals, such sea lions, recuperate on them. Clams, periwinkles, crabs, shrimp, starfish and sea urchins are found on most beaches.

Sand is a sediment made from small grains or particles with diameters between about 60 μm and 2 mm. Mud is a sediment made from particles finer than sand. This small particle size means that mud particles tend to stick together, whereas sand particles do not. Mud is not easily shifted by waves and currents, and when it dries out, cakes into a solid. By contrast, sand is easily shifted by waves and currents, and when sand dries out it can be blown in the wind, accumulating into shifting sand dunes. Beyond the high tide mark, if the beach is low-lying, the wind can form rolling hills of sand dunes. Small dunes shift and reshape under the influence of the wind while larger dunes stabilise the sand with vegetation.

Ocean processes grade loose sediments to particle sizes other than sand, such as gravel or cobbles. Waves breaking on a beach can leave a berm, which is a raised ridge of coarser pebbles or sand, at the high tide mark. Shingle beaches are made of particles larger than sand, such as cobbles, or small stones. These beaches make poor habitats. Little life survives because the stones are churned and pounded together by waves and currents.

Rocky Shores

The relative solidity of rocky shores seems to give them a permanence compared to the shifting nature of sandy shores. This apparent stability is not real over even quite short geological time scales, but it is real enough over the short life of an organism. In contrast to sandy shores, plants and animals can anchor themselves to the rocks.

Tidepools on rocky shores make turbulent habitats for many forms of marine life

Competition can develop for the rocky spaces. For example, barnacles can compete successfully on open intertidal rock faces to the point where the rock surface is covered with them. Barnacles resist desiccation and grip well to exposed rock faces. However, in the crevices of the same rocks, the inhabitants are different. Here mussels can be the successful species, secured to the rock with their byssal threads.

Rocky and sandy coasts are vulnerable because humans find them attractive and want to live near them. An increasing proportion of the humans live by the coast, putting pressure on coastal habitats.

Mudflats

Mudflats become temporary habitats for migrating birds

Mudflats are coastal wetlands that form when mud is deposited by tides or rivers. They are found in sheltered areas such as bays, bayous, lagoons, and estuaries. Mudflats may be viewed geologically as exposed layers of bay mud, resulting from deposition of estuarine silts, clays and marine animal detritus. Most of the sediment within a mudflat is within the intertidal zone, and thus the flat is submerged and exposed approximately twice daily.

Mudflats are typically important regions for wildlife, supporting a large population, although levels of biodiversity are not particularly high. They are of particular importance to migratory birds. In the United Kingdom mudflats have been classified as a Biodiversity Action Plan priority habitat.

Mangrove and Salt Marshes

Mangroves provide nurseries for fish

Mangrove swamps and salt marshes form important coastal habitats in tropical and temperate areas respectively.

Mangroves are species of shrubs and medium size trees that grow in saline coastal sediment habitats in the tropics and subtropics – mainly between latitudes 25° N and 25° S. The saline conditions tolerated by various species range from brackish water, through pure seawater (30 to 40 ppt), to water concentrated by evaporation to over twice the salinity of ocean seawater (up to 90 ppt). There are many mangrove species, not all closely related. The term "mangrove" is used generally to cover all of these species, and it can be used narrowly to cover just mangrove trees of the genus *Rhizophora*.

Mangroves form a distinct characteristic saline woodland or shrubland habitat, called a *mangrove swamp* or *mangrove forest*'. Mangrove swamps are found in depositional coastal environments, where fine sediments (often with high organic content) collect in areas protected from high-energy wave action. Mangroves dominate three quarters of tropical coastlines.

Estuaries

Estuaries occur when rivers flow into a coastal bay or inlet.
They are nutrient rich and have a transition zone which moves from freshwater to saltwater.

An estuary is a partly enclosed coastal body of water with one or more rivers or streams flowing into it, and with a free connection to the open sea. Estuaries form a transition zone between river environments and ocean environments and are subject to both marine influences, such as tides, waves, and the influx of saline water; and riverine influences, such as flows of fresh water and sediment. The inflow of both seawater and freshwater provide high levels of nutrients in both the water column and sediment, making estuaries among the most productive natural habitats in the world.

Most estuaries were formed by the flooding of river-eroded or glacially scoured valleys when sea level began to rise about 10,000-12,000 years ago. They are amongst the most heavily populated areas throughout the world, with about 60% of the world's population living along estuaries and the coast. As a result, estuaries are suffering degradation by many factors, including sedimentation from soil erosion from deforestation; overgrazing and other poor farming practices; overfishing; drainage and filling of wetlands; eutrophication due to excessive nutrients from sewage and animal wastes; pollutants including heavy metals, PCBs, radionuclides and hydrocarbons from sewage inputs; and diking or damming for flood control or water diversion.

Estuaries provide habitats for a large number of organisms and support very high productivity. Estuaries provide habitats for salmon and sea trout nurseries, as well as migratory bird populations. Two of the main characteristics of estuarine life are the variability in salinity and sedimentation. Many species of fish and invertebrates have various methods to control or conform to the shifts in salt concentrations and are termed osmoconformers and osmoregulators. Many animals also burrow to avoid predation and to live in the more stable sedimental environment. However, large numbers of bacteria are found within the sediment which have a very high oxygen demand. This reduces the levels of oxygen within the sediment often resulting in partially anoxic conditions, which can be further exacerbated by limited water flux. Phytoplankton are key primary producers in estuaries. They move with the water bodies and can be flushed in and out with the tides. Their productivity is largely dependent on the turbidity of the water. The main phytoplankton present are diatoms and dinoflagellates which are abundant in the sediment.

Kelp Forests

Kelp forests provide habitat for many marine organisms

Kelp forests are underwater areas with a high density of kelp. They form some of the most productive and dynamic ecosystems on Earth. Smaller areas of anchored kelp are called *kelp beds*. Kelp forests occur worldwide throughout temperate and polar coastal oceans.

Kelp forests provide a unique three-dimensional habitat for marine organisms and are a source for understanding many ecological processes. Over the last century, they have been the focus of extensive research, particularly in trophic ecology, and continue to provoke important ideas that are relevant beyond this unique ecosystem. For example, kelp forests can influence coastal oceanographic patterns and provide many ecosystem services.

However, humans have contributed to kelp forest degradation. Of particular concern are the effects of overfishing nearshore ecosystems, which can release herbivores from their normal population regulation and result in the over-grazing of kelp and other algae. This can rapidly result in transitions to barren landscapes where relatively few species persist.

Frequently considered an ecosystem engineer, kelp provides a physical substrate and habitat for kelp forest communities. In algae (Kingdom: Protista), the body of an individual organism is known as a thallus rather than as a plant (Kingdom: Plantae). The morphological structure of a kelp thallus is defined by three basic structural units:

- The holdfast is a root-like mass that anchors the thallus to the sea floor, though unlike true roots it is not responsible for absorbing and delivering nutrients to the rest of the thallus;
- The stipe is analogous to a plant stalk, extending vertically from the holdfast and providing a support framework for other morphological features;
- The fronds are leaf- or blade-like attachments extending from the stipe, sometimes along its full length, and are the sites of nutrient uptake and photosynthetic activity.

In addition, many kelp species have pneumatocysts, or gas-filled bladders, usually located at the base of fronds near the stipe. These structures provide the necessary buoyancy for kelp to maintain an upright position in the water column.

The environmental factors necessary for kelp to survive include hard substrate (usually rock), high nutrients (e.g., nitrogen, phosphorus), and light (minimum annual irradiance dose $> 50\ E\ m^{-2}$). Especially productive kelp forests tend to be associated with areas of significant oceanographic upwelling, a process that delivers cool nutrient-rich water from depth to the ocean's mixed surface layer. Water flow and turbulence facilitate nutrient assimilation across kelp fronds throughout the water column. Water clarity affects the depth to which sufficient light can be transmitted. In ideal conditions, giant kelp (*Macrocystis spp.*) can grow as much as 30-60 centimetres vertically per day. Some species such as *Nereocystis* are annual while others like *Eisenia* are perennial, living for more than 20 years. In perennial kelp forests, maximum growth rates occur during upwelling months (typically spring and summer) and die-backs correspond to reduced nutrient availability, shorter photoperiods and increased storm frequency.

Seagrass Meadows

Seagrasses are flowering plants from one of four plant families which grow in marine environments. They are called *seagrasses* because the leaves are long and narrow and are very often green, and because the plants often grow in large meadows which look like grassland. Since seagrasses photosynthesize and are submerged, they must grow submerged in the photic zone, where there is enough sunlight. For this reason, most occur in shallow and sheltered coastal waters anchored in sand or mud bottoms.

White-spotted puffers like living in seagrass areas

Seagrasses form extensive beds or meadows, which can be either monospecific (made up of one species) or multispecific (where more than one species co-exist). Seagrass beds make highly diverse and productive ecosystems. They are home to phyla such as juvenile and adult fish, epiphytic and free-living macroalgae and microalgae, mollusks, bristle worms, and nematodes. Few species were originally considered to feed directly on seagrass leaves (partly because of their low nutritional content), but scientific reviews and improved working methods have shown that seagrass herbivory is a highly important link in the food chain, with hundreds of species feeding on seagrasses worldwide, including green turtles, dugongs, manatees, fish, geese, swans, sea urchins and crabs.

Seagrasses are ecosystem engineers in the sense that they partly create their own habitat. The leaves slow down water-currents increasing sedimentation, and the seagrass roots and rhizomes stabilize the seabed. Their importance to associated species is mainly due to provision of shelter (through their three-dimensional structure in the water column), and due to their extraordinarily high rate of primary production. As a result, seagrasses provide coastal zones with ecosystem services, such as fishing grounds, wave protection, oxygen production and protection against coastal erosion. Seagrass meadows account for 15% of the ocean's total carbon storage.

Coral Reefs

Reefs comprise some of the densest and most diverse habitats in the world. The best-known types of reefs are tropical coral reefs which exist in most tropical waters; however, reefs can also exist in cold water. Reefs are built up by corals and other calcium-depositing animals, usually on top of a rocky outcrop on the ocean floor. Reefs can also grow on other surfaces, which has made it possible to create artificial reefs. Coral reefs also support a huge community of life, including the corals themselves, their symbiotic zooxanthellae, tropical fish and many other organisms.

Much attention in marine biology is focused on coral reefs and the El Niño weather phenomenon. In 1998, coral reefs experienced the most severe mass bleaching events on record, when vast expanses of reefs across the world died because sea surface temperatures rose well above normal. Some reefs are recovering, but scientists say that between 50% and 70% of the world's coral reefs are now endangered and predict that global warming could exacerbate this trend.

Open Ocean

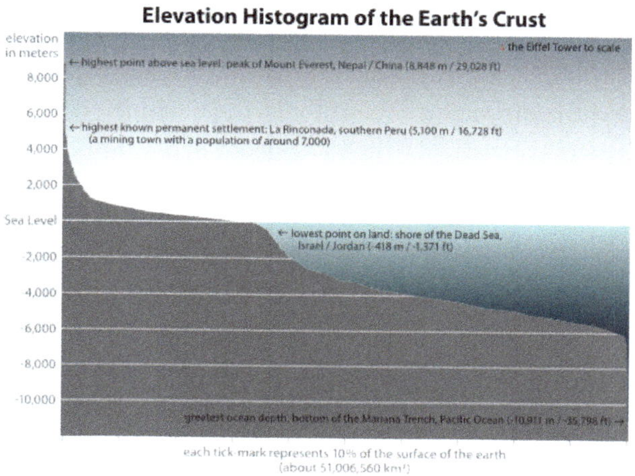

Elevation-area graph showing the proportion of land area at given heights and the proportion of ocean area at given depths

The open ocean is relatively unproductive because of a lack of nutrients, yet because it is so vast, it has more overall primary production than any other marine habitat. Only about 10 percent of marine species live in the open ocean. But among them are the largest and fastest of all marine animals, as well as the animals that dive the deepest and migrate the longest. In the depths lurk animal that, to our eyes, appear hugely alien.

Surface Waters

In the open ocean, sunlit surface epipelagic waters get enough light for photosynthesis, but there are often not enough nutrients. As a result, large areas contain little life apart from migrating animals.

The surface waters are sunlit. The waters down to about 200 metres are said to be in the epipelagic zone. Enough sunlight enters the epipelagic zone to allow photosynthesis by phytoplankton. The epipelagic zone is usually low in nutrients. This partially because the organic debris produced in the zone, such as excrement and dead animals, sink to the depths and are lost to the upper zone. Photosynthesis can happen only if both sunlight and nutrients are present.

In some places, like at the edge of continental shelves, nutrients can upwell from the ocean depth, or land runoff can be distributed by storms and ocean currents. In these areas, given that both sunlight and nutrients are now present, phytoplankton can rapidly establish itself, multiplying so fast that the water turns green from the chlorophyll, resulting in an algal bloom. These nutrient rich surface waters are among the most biologically productive in the world, supporting billions of tonnes of biomass.

"Phytoplankton are eaten by zooplankton - small animals which, like phytoplankton, drift in the ocean currents. The most abundant zooplankton species are copepods and krill: tiny crustaceans that are the most numerous animals on Earth. Other types of zooplankton include jelly fish and the larvae of fish, marine worms, starfish, and other marine organisms". In turn, the zooplankton are eaten by filter-feeding animals, including some seabirds, small forage fish like herrings and sardines, whale sharks, manta rays, and the largest animal in the world, the blue whale. Yet again, moving up the foodchain, the small forage fish are in turn eaten by larger predators, such as tuna, marlin, sharks, large squid, seabirds, dolphins, and toothed whales.

Deep Sea

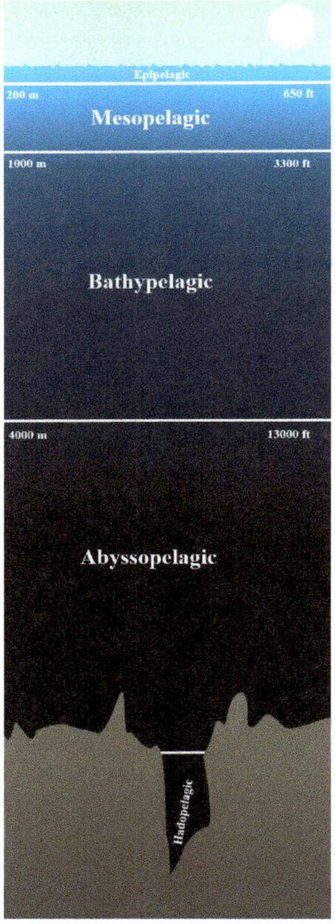

Scale diagram of the layers of the pelagic zone

The deep sea starts at the aphotic zone, the point where sunlight loses most of its energy in the water. Many life forms that live at these depths have the ability to create their own light a unique evolution known as bio-luminescence.

In the deep ocean, the waters extend far below the epipelagic zone, and support very different types of pelagic life forms adapted to living in these deeper zones.

Much of the aphotic zone's energy is supplied by the open ocean in the form of detritus. In deep water, marine snow is a continuous shower of mostly organic detritus falling from the upper layers of the water column. Its origin lies in activities within the productive photic zone. Marine snow includes dead or dying plankton, protists (diatoms), fecal matter, sand, soot and other inorganic dust. The "snowflakes" grow over time and may reach several centimetres in diameter, travelling for weeks before reaching the ocean floor. However, most organic components of marine snow are consumed by microbes, zooplankton and other filter-feeding animals within the first 1,000 metres of their journey, that is, within the epipelagic zone. In this way marine snow may be considered the foundation of deep-sea mesopelagic and benthic ecosystems: As sunlight cannot reach them, deep-sea organisms rely heavily on marine snow as an energy source.

Some deep-sea pelagic groups, such as the lanternfish, ridgehead, marine hatchetfish, and lightfish families are sometimes termed *pseudoceanic* because, rather than having an even distribution in open water, they occur in significantly higher abundances around structural oases, notably seamounts and over continental slopes. The phenomenon is explained by the likewise abundance of prey species which are also attracted to the structures.

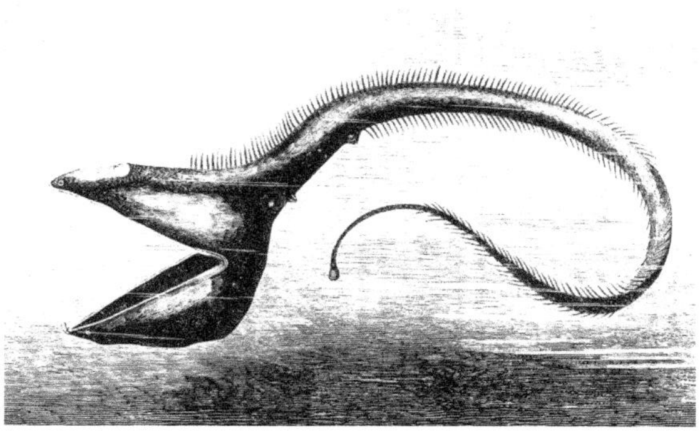

The umbrella mouth gulper eel can swallow a fish much larger than itself

The fish in the different pelagic and deep water benthic zones are physically structured, and behave in ways, that differ markedly from each other. Groups of coexisting species within each zone all seem to operate in similar ways, such as the small mesopelagic vertically migrating plankton-feeders, the bathypelagic anglerfishes, and the deep water benthic rattails.

Ray finned species, with spiny fins, are rare among deep sea fishes, which suggests that deep sea fish are ancient and so well adapted to their environment that invasions by more modern fishes have been unsuccessful. The few ray fins that do exist are mainly in the Beryciformes and Lampriformes, which are also ancient forms. Most deep sea pelagic fishes belong to their own orders, suggesting a long evolution in deep sea environments. In contrast, deep water benthic species, are in orders that include many related shallow water fishes.

The umbrella mouth gulper is a deep sea eel with an enormous loosely hinged mouth. It can open its mouth wide enough to swallow a fish much larger than itself, and then expand its stomach to accommodate its catch.

Sea Floor

Vents and Seeps

Zooarium chimney provides a habitat for vent biota

Hydrothermal vents along the mid-ocean ridge spreading centers act as oases, as do their opposites, cold seeps. Such places support unique marine biomes and many new marine microorganisms and other lifeforms have been discovered at these locations.

Trenches

The deepest recorded oceanic trenches measure to date is the Mariana Trench, near the Philippines, in the Pacific Ocean at 10,924 m (35,838 ft). At such depths, water pressure is extreme and there is no sunlight, but some life still exists. A white flatfish, a shrimp and a jellyfish were seen by the American crew of the bathyscaphe *Trieste* when it dove to the bottom in 1960.

Seamounts

Marine life also flourishes around seamounts that rise from the depths, where fish and other sea life congregate to spawn and feed.

Marsh

A marsh along the edge of a small river

A marsh is a wetland that is dominated by herbaceous rather than woody plant species. Marshes can often be found at the edges of lakes and streams, where they form a transition between the aquatic and terrestrial ecosystems. They are often dominated by grasses, rushes or reeds. If woody plants are present they tend to be low-growing shrubs. This form of vegetation is what differentiates marshes from other types of wetland such as swamps, which are dominated by trees, and mires, which are wetlands that have accumulated deposits of acidic peat.

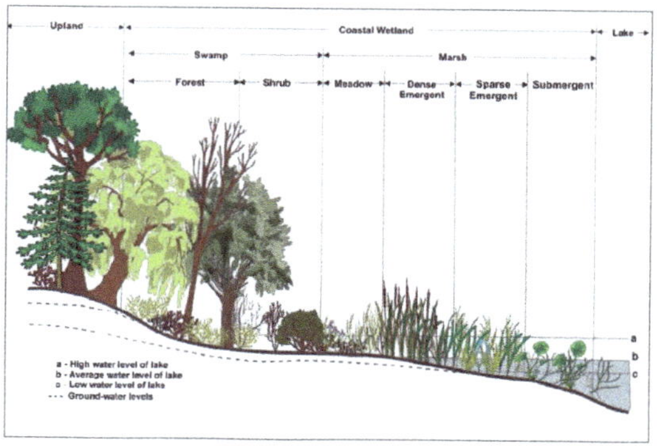

Marsh in shallow water on a lakeshore

Green Cay Wetlands, Palm Beach County, Florida

Basic Information

White water lilies are a typical marsh plant in European areas of deeper water.

Marshes provide a habitat for many species of plants, animals, and insects that have adapted to living in flooded conditions. The plants must be able to survive in wet mud with low oxygen levels.

Many of these plants therefore have aerenchyma, channels within the stem that allow air to move from the leaves into the rooting zone. Marsh plants also tend to have rhizomes for underground storage and reproduction. Familiar examples include cattails, sedges, papyrus and sawgrass. Aquatic animals, from fish to salamanders, are generally able to live with a low amount of oxygen in the water. Some can obtain oxygen from the air instead, while others can live indefinitely in conditions of low oxygen. Marshes provide habitats for many kinds of invertebrates, fish, amphibians, waterfowl and aquatic mammals. Marshes have extremely high levels of biological production, some of the highest in the world, and therefore are important in supporting fisheries. Marshes also improve water quality by acting as a sink to filter pollutants and sediment from the water that flows through them. Marshes (and other wetlands) are able to absorb water during periods of heavy rainfall and slowly release it into waterways and therefore reduce the magnitude of flooding. The pH in marshes tends to be neutral to alkaline, as opposed to bogs, where peat accumulates under more acid conditions.

Many kinds of birds nest in marshes; this one is a yellow-headed blackbird.

Types of Marshes

A salt marsh in Scotland

Marshes differ depending mainly on their location and salinity. Both of these factors greatly influence the range and scope of animal and plant life that can survive and reproduce in these environments. The three main types of marsh are salt marshes, freshwater tidal marshes, and freshwater marshes. These three can be found worldwide and each contains a different set of organisms.

Salt Marshes

Saltwater marshes are found around the world in mid to high latitudes, wherever there are sections of protected coastline. They are located close enough to the shoreline that the motion of the tides affects them, and, sporadically, they are covered with water. They flourish where the rate of sediment buildup is greater than the rate at which the land level is sinking. Salt marshes are dominated by specially adapted rooted vegetation, primarily salt-tolerant grasses.

Salt marshes are most commonly found in lagoons, estuaries, and on the sheltered side of shingle or sandspit. The currents there carry the fine particles around to the quiet side of the spit and sediment begins to build up. These locations allow the marshes to absorb the excess nutrients from the water running through them before they reach the oceans and estuaries. These marshes are slowly declining. Coastal development and urban sprawl has caused significant loss of these essential habitats.

Freshwater Tidal Marshes

Although considered a freshwater marsh, this form of marsh is affected by the ocean tides. However, without the stresses of salinity at work in its saltwater counterpart, the diversity of the plants and animals that live in and use freshwater tidal marshes is much higher than in salt marshes. The most serious threats to this form of marsh are the increasing size and pollution of the cities surrounding them.

Freshwater Marshes

A wet meadow adjacent to Big Bear Lake, San Bernardino Mountains, California

Ranging greatly in both size and geographic location, freshwater marshes make up the most common form of wetland in North America. They are also the most diverse of the three types of marsh. Some examples of freshwater marsh types in North America are:

Wet Meadows

Wet meadows occur in areas such as shallow lake basins, low-lying depressions, and the land between shallow marshes and upland areas. They also occur on the edges of large lakes and rivers. Wet meadows often have very high plant diversity and high densities of buried seeds. They are regularly flooded but are often dry in the summer.

Vernal Pools

Vernal pools are a type of marsh found only seasonally in shallow depressions in the land. They can be covered in shallow water, but in the summer and fall, they can be completely dry. In western North America, vernal pools tend to form in open grasslands, whereas in the east they often occur in forested landscapes. Further south, vernal pools form in pine savannas and flatwoods. Many amphibian species depend upon vernal pools for spring breeding; these ponds provide habitat free from fish which eat the eggs and young of amphibians. An example is the endangered gopher frog (*Rana sevosa*). Similar temporary ponds occur in other world ecosystems, where they may have local names. However, the term vernal pool can be applied to all such temporary pool ecosystems.

Playa Lakes

Playa lakes are a form of shallow freshwater marsh that occurs in the southern high plains of the United States. Like vernal pools, they are only present at certain times of the year and generally have a circular shape. As the playa dries during the summer, conspicuous plant zonation develops along the shoreline.

Prairie Potholes

Aerial view of prairie potholes

Prairie potholes are found in the northern parts of North America as the Prairie Pothole Region. These landscapes were once covered by glaciers, and as a result shallow depressions were formed in great numbers. These depressions fill with water in the spring. They provide important breeding habitats for many species of waterfowl. Some pools only occur seasonally while others retain enough water to be present all year.

Riverine Wetlands

Many kinds of marsh occur along the fringes of large rivers. The different types are produced by factors such as water level, nutrients, ice scour, and waves.

Restoration

Some areas of the world have already lost 90% of their wetlands, including marshes. They have been drained to create agricultural land or filled to accommodate urban sprawl. Restoration is the process of returning marshes to the landscape to replace those lost in the past. Restoration can be done on a large scale, such as by allowing rivers to flood naturally in the spring, or on a small scale by returning wetlands to urban landscapes.

Halophyte

Spartina alterniflora (cordgrass), a halophyte.

A halophyte is a plant that grows in waters of high salinity, coming into contact with saline water through its roots or by salt spray, such as in saline semi-deserts, mangrove swamps, marshes and sloughs and seashores. An example of a halophyte is the salt marsh grass *Spartina alterniflora* (smooth cordgrass). Relatively few plant species are halophytes—perhaps only 2% of all plant species.

The large majority of plant species are glycophytes, which are not salt-tolerant and are damaged fairly easily by high salinity.

Salt Tolerance

One quantitative measure of salt tolerance (halotolerance) is the total dissolved solids in irrigation water that a plant can tolerate. Seawater typically contains 40 grams per litre (g/l) of dissolved salts (mostly sodium chloride). Beans and rice can tolerate about 1–3 g/l, and are considered glycophytes (as are most crop plants). At the other extreme, *Salicornia bigelovii* (dwarf glasswort) grows well at 70 g/l of dissolved solids, and is a promising halophyte for use as a crop. Plants such as barley (*Hordeum vulgare*) and the date palm (*Phoenix dactylifera*) can tolerate about 5 g/l, and can be considered as marginal halophytes.

Adaptation to saline environments by halophytes may take the form of salt tolerance or salt avoidance. Plants that avoid the effects of high salt even though they live in a saline environment may be referred to as facultative halophytes rather than 'true', or obligatory, halophytes.

For example, a short-lived plant species that completes its reproductive life cycle during periods (such as a rainy season) when the salt concentration is low would be avoiding salt rather than tolerating it. Or a plant species may maintain a 'normal' internal salt concentration by excreting excess salts through its leaves, by way of a hydathode, or by concentrating salts in leaves that later die and drop off. In an effort to improve agricultural production in regions where crops are exposed to salinity, research is focused on improving understanding of the various mechanisms whereby plants respond to salinity stress, so that more robust crop halophytes may be developed.

Adaptive responses to salinity stress have been identified at molecular, cellular, metabolic, and physiological levels.

Examples

Some halophytes are:

- *Anemopsis californica* (*yerba mansa*, lizard tail)
- *Atriplex* (saltbush, orache, orach)
- *Attalea speciosa* (babassu)
- *Panicum virgatum* (switchgrass)
- *Salicornia bigelovii* (dwarf glasswort, pickleweed)
- *Spartina alterniflora* (smooth cordgrass)
- *Tetragonia tetragonoides* (warrigal greens, *kōkihi*, sea spinach)

As Biofuel

Some halophytes are being studied for use as "3rd-generation" biofuel precursors. Halophytes such as *Salicornia bigelovii* can be grown in harsh environments and typically do not compete with food crops for resources, making them promising sources of biodiesel or bioalcohol.

References

- Browne, R. A. (1981). "Lakes as islands: biogeographic distribution, turnover rates, and species composition in the lakes of central New York". Journal of Biogeography. 8 1: 75–83. JSTOR 2844594. doi:10.2307/2844594
- Giller, S.; B. Malmqvist (1998). The Biology of Streams and Rivers. Oxford University Press, Oxford. p. 296. ISBN 0198549776
- Pidwirny, Michael (2006). "Fundamentals of Physical Geography (2nd Edition)". PhysicalGeography.net. Retrieved 2011-04-19
- Hillebrand, H.; A. I. Azovsky (2001). "Body size determines the strength of the latitudinal diversity gradient". Ecography. 24 (3): 251–256. doi:10.1034/j.1600-0587.2001.240302.x
- Keddy, P.A. and A. A. Reznicek. 1986. Great Lakes vegetation dynamics: the role of fluctuating water levels and buried seeds. Journal of Great Lakes Research 12: 25-36
- Moss, B. (1998). Ecology of Freshwaters: man and medium, past to future. Blackwell Science, London. p. 557. ISBN 0632035129
- Glenn, Edward P.; Brown, J. Jed; O'Leary, James W. (August 1998). "Irrigating Crops with Seawater" (PDF). Scientific American. USA: Scientific American, Inc. (August 1998): 76–81 [79]. Retrieved 2008-11-17
- Glenn, E. P.; et al. (1999). "Salt tolerance and crop potential of halophytes". Critical Review in Plant Sciences. 18 (2): 227–55. doi:10.1080/07352689991309207
- Gupta, Bhaskar; Huang, Bingru (3 April 2014). "Mechanism of Salinity Tolerance in Plants: Physiological, Biochemical, and Molecular Characterization". International Journal of Genomics. doi:10.1155/2014/701596
- Nybakken JW and Bertness MD (2005) Marine biology: an ecological approach Sixth edition, Pearson/Benjamin Cummings. ISBN 978-0-8053-4582-7 – organized by habitat, not classification

2

An Overview of Aquatic Plants

The plants found in the water bodies are known as aquatic plants. Equisetum fluviatile, hippuris vulgaris, sagittaria, ceratophyllum and seagrass are some of the types of aquatic plants. This chapter is an overview of the subject matter incorporating all the major aspects of aquatic plants.

Aquatic Plant

The flower of *Nymphaea alba*, a species of water lily

Aquatic plants are plants that have adapted to living in aquatic environments (saltwater or freshwater). They are also referred to as hydrophytes or macrophytes. These plants require special adaptations for living submerged in water, or at the water's surface. The most common adaptation is aerenchyma, but floating leaves and finely dissected leaves are also common. Aquatic plants can only grow in water or in soil that is permanently saturated with water. They are therefore a common component of wetlands.

The principal factor controlling the distribution of aquatic plants is the depth and duration of flooding. However, other factors may also control their distribution, abundance, and growth form, including nutrients, disturbance from waves, grazing, and salinity.

Aquatic vascular plants have originated on multiple occasions in different plant families; they can be ferns or angiosperms (including both monocots and dicots). Seaweeds are not vascular plants; rather they are multicellular marine algae, and therefore are not typically included among aquatic plants. A few aquatic plants are able to survive in brackish, saline, and salt water. Examples are found in genera such as *Thalassia* and *Zostera*. Although most aquatic plants can reproduce by flowering and setting seed, many also have extensive asexual reproduction by means of rhizomes, turions, and fragments in general.

One of the largest aquatic plants in the world is the Amazon water lily; one of the smallest is the minute duckweed. Many small aquatic animals use plants like duckweed for a home, or for protection from predators, but areas with more vegetation are likely to have more predators. Some other familiar examples of aquatic plants might include floating heart, water lily, lotus, and water hyacinth.

Some aquatic plants are used by humans as a food source. Examples include wild rice (*Zizania*), water caltrop (*Trapa natans*), Chinese water chestnut (*Eleocharis dulcis*), Indian lotus (*Nelumbo nucifera*), water spinach (*Ipomoea aquatica*), and watercress (*Rorippa nasturtium-aquaticum*).

Morphological Classification

Water lilies grow rooted in the bottom with leaves that float on the water surface.

The many possible classifications of aquatic plants are based upon morphology. One example has six groups as follows:

- Amphiphytes: plants that are adapted to live either submerged or on land
- Elodeids: stem plants that complete their entire lifecycle submerged, or with only their flowers above the waterline
- Isoetids: rosette plants that complete their entire lifecycle submerged
- Helophytes: plants rooted in the bottom, but with leaves above the waterline
- Nymphaeids: plants rooted in the bottom, but with leaves floating on the water surface
- Pleuston: vascular plants that float freely in the water

Ceratophyllum submersum, a free-floating plant that grows completely submerged	*Pistia stratiotes*, an example of a pleuston, a plant that floats freely on the water surface	*Lysichiton americanus* grows rooted in the bottom with leaves and flowers above the waterline.

Equisetum Fluviatile

Equisetum fluviatile, the water horsetail or swamp horsetail, is a vascular plant that commonly grows in dense colonies along freshwater shorelines or in shallow water in ponds, swamps, ditches, and other sluggish or still waters with mud bottoms. It is a perennial herbaceous species, growing 30–100 cm (rarely 140 cm) tall with erect dark green stems 2–8 mm in diameter, smooth, with about 10–30 fine ridges. At each joint, the stem has a whorl of tiny, black-tipped scale leaves 5–10 mm long. Many, but not all, stems also have whorls of short ascending and spreading branches 1–5 cm long, with the longest branches on the lower middle of the stem. The side branches are slender, dark green, and have 1–8 nodes with a whorl of five scale leaves at each node. The water horsetail has the largest central hollow of the horsetails, with 80% of the stem diameter typically being hollow.

The stems readily pull apart at the joints, and both fertile and sterile stems look alike.

The water horsetail reproduces both by spores and vegetatively by rhizomes. It primarily reproduces by vegetative means, with the majority of shoots arising from rhizomes. Spores are produced in blunt-tipped cones at the tips of some stems. The spore cones are yellowish-green, 1-2 cm long and 1 cm broad, with numerous scales in dense whorls.

The water horsetail ranges throughout the temperate Northern Hemisphere, from Eurasia south to central Spain, northern Italy, the Caucasus, China, Korea and Japan, and in North America from the Aleutian Islands to Newfoundland, south to Oregon, Idaho, northwest Montana, northeast Wyoming, West Virginia and Virginia.

This horsetail is sometimes seen as an invasive species because it is very hardy and tends to overwhelm other garden plants unless it is contained. When planting, it is best to plant them with the rhizome in a container.

The water horsetail is most often confused with the marsh horsetail *E. palustre*, which has rougher stems with fewer (4-8) stem ridges with a smaller hollow in the stem centre, and longer spore cones 2–4 cm long.

Uses

The water horsetail has historically been used by both Europeans and Native Americans for scouring, sanding, and filing because of the high silica content in the stems. Early spring shoots were eaten. Medically it was used by the ancient Greeks and Romans to stop bleeding and treat kidney ailments, ulcers, and tuberculosis, and by the ancient Chinese to treat superficial visual obstructions. Rootstocks and stems are sometimes eaten by waterfowl. Horsetails absorb heavy metals from the soil, and are often used in bioassays for metals.

According to Carl Linnaeus, reindeer, which refuse ordinary hay, will eat this horsetail, which is juicy, and that it is cut as fodder in the north of Sweden for cows, with a view to increasing their milk yield, but that horses will not touch it.

Hippuris Vulgaris

Hippuris vulgaris (from Greek *horse* and *tail*), known as mare's-tail or common mare's tail, is a common aquatic plant of Eurasia and North America ranging from Greenland to the Tibetan

Plateau to Arizona. It prefers non-acidic waters. The species is also sometimes called horsetail, a name which is better reserved to the horsetails of genus *Equisetum*. These are unrelated to *H. vulgaris*, although there is some resemblance in appearance.

Description

The common mare's tail is a creeping, perennial herb, found in shallow waters and mud flats. It roots underwater, but most of its leaves are above the water surface. The leaves occur in whorls of 6-12; those above water are 0.5 to 2.5 cm long and up to 3 mm wide, whereas those under water are thinner and limper, and longer than those above water, especially in deeper streams. The stems are solid and unbranched but often curve, and can be up to 60 cm long. In shallow water they project 20–30 cm out of the water. It grows from stout rhizomes. The flowers are inconspicuous, and not all plants produce them. Studies of *H. vulgaris* in the Tibetan Plateau have shown that it is a prolific methane emitter. *H. vulgaris's* roots extend into the anoxic zone of wetland soils and create a conduit for methane produced in the anoxic zone to travel to the atmosphere.

Uses

In herbal medicine, mare's tail has a number of uses, chiefly to do with healing wounds, e.g. stopping internal and external bleeding, curing stomach ulcers, and soothing inflammation of the skin. It can also be a troublesome weed, obstructing the flow of water in rivers and ditches.

Sagittaria

Sagittaria is a genus of about 30 species of aquatic plants whose members go by a variety of common names, including arrowhead, duck potato, katniss, *Omodaka* (沢瀉 in Japanese), swamp potato, tule potato, and wapato (or wapatoo). Most are native to South, Central, and North America, but there are also some from Europe, Africa, and Asia.

Sagittaria plant stock (the perennial rhizome) is a horizontal creeper (stoloniferous) and obliquely obovate, the margins winged, with apical or ventral beak; in other words, they are a small, dry, one-seeded fruit that do not open to release the seed, set on a slant, narrower at the base, with winged edges, and having a "beaked" aperture (one side longer than the other) for sprouting, set above or below the fruit body.

Sagittaria sagittifolia flowers

One of the names for this plant is derived from the edible underwater tuber that the plant produces. In late fall or early spring, disturbing the aquatic mud in which the plant grows will cause its small tubers to float to the surface where they can be harvested, and then boiled.

Omodaka

Species

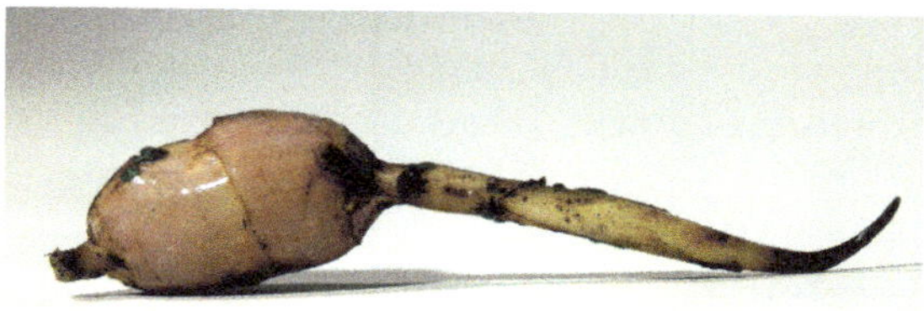

Wapato bulb

Crisps (chips) made from tuber

Accepted species:

1. *Sagittaria aginashii* Makino - Japan, Korea, Primorye
2. *Sagittaria ambigua* J. G. Sm. – Missouri Arrowhead - from Oklahoma to Indiana
3. *Sagittaria australis* (J. G. Sm.) Small – Appalachian Arrowhead - southeastern US from Louisiana to Florida and as far north as Iowa and New Jersey

4. *Sagittaria brevirostra* Mack. & Bush – Shortbeak Arrowhead - central US (Great Plains, Mississippi and Ohio Valleys, Great Lakes); also Virginia and Saskatchewan; naturalized in California
5. *Sagittaria chapmanii* (J.G.Sm.) C.Mohr - from Texas to the Carolinas
6. *Sagittaria cristata* Engelm – Crystal arrowhead - Great Lakes region
7. *Sagittaria cuneata* E.P.Sheld. – Wapato, Northern Arrowhead, Swamp Potato - most of Canada including Yukon and Northwest Territories; Alaska; western and northeastern USA
8. *Sagittaria demersa* J.G.Sm. – Chihuahuan arrowhead - New Mexico, northeastern Mexico
9. *Sagittaria engelmanniana* J.G.Sm. – Engelmann's arrowhead - eastern US from Mississippi to Vermont
10. *Sagittaria fasciculata* E.O.Beal – Bunched Arrowhead - North and South Carolina
11. *Sagittaria filiformis* J.G.Sm. – Threadleaf Arrowhead - eastern US from Alabama to Maine
12. *Sagittaria graminea* Michx. – Grassy Arrowhead, Grass-leaved Arrowhead - Cuba; much of eastern and central USA; eastern Canada; naturalized in Washington State and in Vietnam
13. *Sagittaria guayanensis* Kunth – Guyanese Arrowhead - widespread across Latin America, the West Indies, China, India, Southeast Asia; introduced into Louisiana
14. *Sagittaria intermedia* Micheli in A.L.P.P.de Candolle & A.C.P.de Candolle - Greater Antilles, Colombia, southern Mexico
15. *Sagittaria isoetiformis* J.G.Sm. – Quillwort Arrowhead - Cuba, from Florida, Georgia, Alabama, Mississippi, Carolinas
16. *Sagittaria kurziana* Glück – Springtape or Strap-leaf Sagittaria - Florida; naturalized in Mariana Islands
17. *Sagittaria lancifolia* L. – Bulltongue Arrowhead - southeastern USA from Texas to Delaware; West Indies; Latin America from southern Mexico to Brazil
18. *Sagittaria latifolia* Willd. – Duck-potato, Broad-leaved Arrowhead, Wapato - widespread across most of North America, the West Indies and northern South America; naturalized in Hawaii, the western Himalayas and parts of Europe
19. *Sagittaria lichuanensis* J.K.Chen, X.Z.Sun & H.Q.Wang - southern China
20. *Sagittaria longiloba* Engelm. ex J.G. Sm. – Longbarb Arrowhead - southern Great Plains, Arizona, New Mexico, California, Mexico, Nicaragua, Venezuela
21. *Sagittaria × lunata* C.D.Preston & Uotila - Sweden, Finland, northern Russia *(S. natans × S. sagittifolia)*

22. *Sagittaria macrocarpa* J.G.Sm. – Large-fruited Arrowhead - North and South Carolina
23. *Sagittaria macrophylla* Zucc. – Papa de agua - Mexico
24. *Sagittaria montevidensis* Cham. & Schltdl. – California Arrowhead – widespread across much of USA, Mexico and South America
25. *Sagittaria natans* Pall. - widespread across northern Europe and Asia from Sweden to Kamchatka; Russia, China, Kazakhstan, Japan, Korea
26. *Sagittaria papillosa* Buchenau – Nipplebract Arrowhead - Texas, Louisiana, Mississippi, Arkansas, Oklahoma
27. *Sagittaria planitiana* G.Agostini - Brazil, Venezuela
28. *Sagittaria platyphylla* (Engelm.) J.G.Sm. – Delta Arrowhead, Delta Duck-potato - south-central USA with scattered populations in southeast, the Ohio Valley and Washington State; also Mexico and Panama; naturalized in South Australia, Italy, Java, Caucasus
29. *Sagittaria potamogetifolia* Merr. - southern China
30. *Sagittaria pygmaea* Miq. – Pygmy arrowhead - China, Japan, Korea, Himalayas, Thailand, Vietnam
31. *Sagittaria rhombifolia* Cham. - Costa Rica; widespread across much of South America
32. *Sagittaria rigida* Pursh. – Canadian Arrowhead - Canada from Quebec to Saskatchewan; common in northeastern and north-central USA from Arkansas and Nebraska east to Virginia and New England; scattered populations in California, Idaho and Washington State; naturalised in Great Britain
33. *Sagittaria sagittifolia* L. – Arrowhead - widespread across most of Europe; Siberia, Caucasus, Turkey
34. *Sagittaria sanfordii* Greene – Valley Arrowhead - endemic to California
35. *Sagittaria secundifolia* Kral – Little River Arrowhead - Georgia and Alabama
36. *Sagittaria sprucei* Micheli in A.L.P.P.de Candolle & A.C.P.de Candolle - Colombia, Peru, Venezuela, northern Brazil
37. *Sagittaria subulata* L. Buchenau – Narrow-leaved Arrowhead - eastern USA from Louisiana to Massachusetts; naturalized in Great Britain, Azores and Java
38. *Sagittaria tengtsungensis* H.Li - Tibet, Nepal, Bhutan, Yunnan
39. *Sagittaria teres* S.Watson – Slender Arrowhead - northeastern USA
40. *Sagittaria trifolia* L. – Threeleaf Arrowhead - widespread across much of Asia including Siberia, China, Japan, India, Iran, Indonesia, Philippines, etc.; also Ukraine and European Russia; naturalized in Fiji and Polynesia

Formerly Placed Here

- *Echinodorus palaefolius* (Nees & Mart.) J. F. Macbr. (as *S. palaefolia* Nees & Mart.)
- *Limnophyton obtusifolium* (L.) Miq. (as *S. obtusifolia* L.)
- *Wiesneria triandra* (Dalzell) Micheli (as *S. triandra* Dalzell)

Zostera

Zostera sp in Mussel Ridge Channel, Birch Island, Maine

Zostera is a small genus of widely distributed seagrasses, commonly called marine eelgrass or simply eelgrass. The genus *Zostera* contains 15 species.

Ecology

Zostera marina is found on sandy substrates or in estuaries, usually submerged or partially floating. Most *Zostera* are perennial. They have long, bright green, ribbon-like leaves, the width of which are about 1 centimetre (0.4 in). Short stems grow up from extensive, white branching rhizomes. The flowers are enclosed in the sheaths of the leaf bases; the fruits are bladdery and can float.

Zostera beds are important for sediment deposition, substrate stabilization, as substrate for epiphytic algae and micro-invertebrates, and as nursery grounds for many species of economically important fish and shellfish. *Zostera* often forms beds in bay mud in the estuarine setting. It is an important food for Brent geese and wigeons, and even (occasionally) caterpillars of the grass moth *Dolicharthria punctalis*.

The slime mold *Labyrinthula zosterae* can cause the wasting disease of *Zostera*, with *Z. marina* being particularly susceptible, causing a decrease in the populations of the fauna that depend on *Zostera*.

Zostera is able to maintain its turgor at a constant pressure in response to fluctuations in environmental osmolarity. It achieves this by losing solutes as the tide goes out and gaining solutes as the tide comes in.

Distribution

The genus as a whole is widespread throughout seashores of much of the Northern Hemisphere as well as Australia, New Zealand, Southeast Asia and southern Africa. The discovery of *Z. chilensis* in 2005 adds an isolated population on the Pacific coast of South America to the distribution. One species (*Z. noltii*) occurs along the land-locked Caspian Sea.

Uses

Eelgrass has been used for food by the Seri tribe of Native Americans on the coast of Sonora, Mexico. The rhizomes and leaf-bases of eelgrass were eaten fresh or dried into cakes for winter food. It was also used for smoking deer meat. The Seri language has many words related to eelgrass and eelgrass-harvesting. The month of April is called *xnoois ihaat iizax*, literally "the month when the eelgrass seed is mature".

Zostera has also been used as packing material and as stuffing for mattresses and cushions.

On the Danish island of Læsø it has been used for thatching roofs. Roofs of eelgrass are said to be heavy, but also much longer-lasting and easier to thatch and maintain than roofs done with more conventional thatching material.

Zostera can also be utilized to produce biomass energy using the Jean Pain method.

Pistia

Pistia is a genus of aquatic plant in the arum family, Araceae. The single species it comprises, Pistia stratiotes, is often called water cabbage, water lettuce, Nile cabbage, or shellflower. Its native distribution is uncertain, but probably pantropical; it was first discovered from the Nile near Lake Victoria in Africa. It is now present, either naturally or through human introduction, in nearly all tropical and subtropical fresh waterways. The genus name is derived from the Greek word (*pistos*), meaning "water," and refers to the aquatic nature of the plants.

Description

19th century illustration of *Pistia stratiotes*

Water lettuce in a home aquarium

It is a perennial monocotyledon with thick, soft leaves that form a rosette. It floats on the surface of the water, its roots hanging submersed beneath floating leaves. The leaves can be up to 14 cm long and have no stem. They are light green, with parallel veins, wavy margins and are covered in short hairs which form basket-like structures which trap air bubbles, increasing the plant's buoyancy. The flowers are dioecious, and are hidden in the middle of the plant amongst the leaves. Small green berries form after successful fertilization. The plant can also undergo asexual reproduction. Mother and daughter plants are connected by a short stolon, forming dense mats.

Rails (Rallidae, "tikling"), feeding in Pistia stratiotes, Paombong, Bulacan).

Ecology

Water lettuce is among the world's most productive freshwater aquatic plants. In waters with high nutrient content, particularly those that have been contaminated with human loading of sewage or fertilizers, water lettuce can often exhibit weedy overgrowth behavior. It may also commonly become weedy in hydrologically altered systems such as canals and reservoirs.

It is a common aquatic plant in the southeast United States, particularly in Florida. It was originally noted in Florida by the explorers John and William Bartram during the period of 1765-1774, and fossil reports in Florida dating to the late-Pleistocene (~12,000 BP) and early Holocene (~3,500 BP) indicate its native presence in southeast North America.

Severe overgrowth of water lettuce can block gas exchange at the air-water interface, reducing the oxygen in the water and killing fish. Large mats can also block light, shade native submerged plants, and alter immersed plant communities by crushing them.

Mosquitoes of the genus *Mansonia* complete their life cycle only in the presence of aquatic plants such as *Pistia*, laying their eggs under the leaves. The emerging larvae fall into the water within 24 hours and stay attached to the *Pistia* root (which is rich with air sacs) with the help of a serrated siphon tube for respiration and develop into pupa. The pupa is also attached to the pistia root with the serrated piercing siphon tube. The egg to adult mosquito development is completed within 7 days.

Control

Pistia can be controlled by mechanical harvesters that remove the water lettuce from the water and transport it to disposal on shore. Aquatic herbicides may also be used. Two insects are also being used as a biological control. Adults and larvae of the South American weevil *Neohydronomous affinis* feed on *Pistia* leaves, as do the larvae of moth *Spodoptera pectinicornis* from Thailand. Both are proving to be useful tools in the management of *Pistia*.

In the Amazon basin, *Pistia* is a food source of freshwater turtles.

Native Range

The center of origin of *Pistia stratiotes* is uncertain. There are global fossil records of it. Although it is depicted in Egyptian hieroglyphics and the ancient Greek botanists Dioscorides and Theophrastus described it from the Nile River, suggesting an African origin, it hosts many coevolved specialist insect herbivores in Brazil and Argentina, rather suggesting a South American origin.

Uses

Water lettuce is often used in tropical aquariums to provide cover for fry and small fish. It is also helpful as it outcompetes algae for nutrients in the water, thereby preventing massive algal blooms.

Ceratophyllum

Ceratophyllum is a cosmopolitan genus of flowering plants including four accepted species in 2016, commonly found in ponds, marshes, and quiet streams in tropical and in temperate regions. It is the only genus in the family Ceratophyllaceae, itself the only family in the order Ceratophyllales. They are usually called coontails or hornworts, although hornwort is also used for unrelated plants of the division Anthocerotophyta.

Ceratophyllum grows completely submerged, usually, though not always, floating on the surface, and does not tolerate drought. The plant stems can reach 1–3 m in length. At intervals along nodes of the stem they produce rings of bright green leaves, which are narrow and often much-branched. The forked leaves are brittle and stiff to the touch in some species, softer in others. The plants have no roots at all, but sometimes they develop modified leaves with a rootlike appearance, which anchor the plant to the bottom. The flowers are small and inconspicuous, with the male and female flowers on the same plant. In ponds it forms thick buds (turions) in the autumn that sink to the bottom which give the impression that it has been killed by the frost but come spring these will grow back into the long stems slowly filling up the pond.

Hornwort plants float in great numbers just under the surface. They offer excellent protection to fish-spawn, but also to snails that are infected with *Bilharzia*. Because of their appearance and their high oxygen production, they are often used in freshwater aquaria.

Taxonomy

Ceratophyllum is considered distinctive enough to warrant its own family, Ceratophyllaceae. It was considered a relative of Nymphaeaceae and included in Nymphaeales in the Cronquist system, but recent research has shown that it is not closely related to Nymphaeaceae or any other extant plant family. Some early molecular phylogenies suggested it was the sister group to all other angiosperms, but more recent research suggests that it is the sister group to the eudicots. The APG III system placed the family in its own order, the Ceratophyllales.

Species

The division of the genus into species is not completely settled. More than 30 species have been described, but many are probably just variants of these more widely accepted species:

- *Ceratophyllum demersum* L. (rigid hornwort or common hornwort) - cosmopolitan
- *Ceratophyllum echinatum* A.Gray (spineless hornwort) - North America
- *Ceratophyllum muricatum* Cham. (prickly hornwort) - widespread in many places though not all countries
- *Ceratophyllum platyacanthum* Cham. - Scattered locations in Germany, Hungary, France, Russia, China, Japan, Korea
- *Ceratophyllum submersum* L. (soft hornwort or tropical hornwort) - Europe, Central Asia, northern Africa, scattered places in tropical Africa, Turkey, Oman, Florida, Dominican Republic

Of these, *Ceratophyllum demersum* is widespread, with a global distribution; the others all have more restricted ranges.

Bacopa

Bacopa is a genus of 70–100 aquatic plants belonging to the family Plantaginaceae. It is commonly known as Waterhyssop (or Water Hyssop, though this is more misleading as *Bacopa* is not very closely related to hyssop but simply has a somewhat similar appearance).

Description

They are annual or perennial, with decumbent or erect stems. The leaves are opposite or whorled, and sessile. The leaf blade is regular, round to linear, and the venation is palmate or pinnate. Its stems are hairy or smooth. The flowers are produced solitary or in pairs from leaf axils, usually radially symmetrical, with five sepals and five petals, and they are usually white, blue, or purple. Dispersal and propagation is by seeds and stem fragments. Crushed leaves have a distinctive 'lemon' scent.

Bacopa monnieri in Hyderabad, India.

Bacopa myriophylloides

Habitat and Range

Bacopa species are found in tropical and subtropical regions of the world, particularly the Americas. A few are regarded as weeds and excess stock should not be dumped in warmer regions. Most grow in moist amphibious conditions, though some like *B. myriophylloides* seem to be wholly aquatic.

Uses

Bacopa monnieri, syn. *Herpestis monniera*, is important in Ayurvedic medicine and has been used, particularly in India, for several thousand years in the belief that it helps mental functioning, including comprehension (Dhi), memory (Dhriti), and recollection (Smriti).

Cultivation

Some of these species are commonly used in freshwater aquariums and around and in ponds in warmer climates. Most are easy to grow and are slow-growing, but require medium to high amounts of light. Most are easy to grow and will tolerate a wide range of water conditions. *B. caroliniana* at least will tolerate brackish water. Plants are usually best planted in groups in the middle to background. Algal infestation can be a problem in brighter lighting condition.

Selected Species

• *Bacopa australis*	• *Bacopa madagascariensis*
• *Bacopa caroliniana* (Walter) B.L.Rob. – lemon bacopa, blue water hyssop, giant bacopa	• *Bacopa monnieri* (L.) Pennell – water hyssop, moneywort, herb of grace
• *Bacopa decumbens* (Fernald) F.N.Williams	• *Bacopa myriophylloides*
• *Bacopa crenata*	• *Bacopa repens* (Sw.) Wettst.
• *Bacopa eisenii* (Kellogg) Pennell – Gila River waterhyssop	• *Bacopa rotundifolia* (Michx.) Wettst. – disc waterhyssop
• *Bacopa innominata* (G.Maza) Alain – tropical waterhyssop	• *Bacopa stricta* (Schrad.) B.L.Robins. – yerba de culebra

Formerly Placed

- *Mecardonia procumbens* (Mill.) Small (as *B. procumbens* (Mill.) Greenm.)

Seagrass

Floridian seagrass bed

Seagrasses are flowering plants (angiosperms) belonging to four families (Posidoniaceae, Zosteraceae, Hydrocharitaceae and Cymodoceaceae), all in the order Alismatales (in the class of monocotyledons), which grow in marine, fully saline environments. There are 12 genera with some 60 species known.

Ecology

These unusual marine flowering plants are called *seagrasses* because in many species the leaves are long and narrow, grow by rhizome extension, and often grow in large “meadows”, which look like grassland: in other words, many of the species of seagrasses superficially resemble terrestrial grasses of the family Poaceae.

Like all autotrophic plants, seagrasses photosynthesize so are limited to growing in the submerged photic zone, and most occur in shallow and sheltered coastal waters anchored in sand or mud bottoms. Most species undergo submarine pollination and complete their entire life cycle underwater.

White-spotted puffers are often found in seagrass areas.

Seagrasses form extensive beds or meadows, which can be either monospecific (made up of a single species) or in mixed beds where more than one species coexist. In temperate areas, usually one or a few species dominate (like the eelgrass *Zostera marina* in the North Atlantic), whereas tropical beds usually are more diverse, with up to thirteen species recorded in the Philippines.

Seagrass beds are highly diverse and productive ecosystems, and can harbor hundreds of associated species from all phyla, for example juvenile and adult fish, epiphytic and free-living macroalgae and microalgae, mollusks, bristle worms, and nematodes. Few species were originally considered to feed directly on seagrass leaves (partly because of their low nutritional content), but scientific reviews and improved working methods have shown that seagrass herbivory is a highly important link in the food chain, with hundreds of species feeding on seagrasses worldwide, including green turtles, dugongs, manatees, fish, geese, swans, sea urchins and crabs.

Some fish species that visit/feed on seagrasses raise their young in adjacent mangroves or coral reefs. Also, seagrasses trap sediment and slow down water movement, causing suspended sediment to fall out. The trapping of sediment benefits coral by reducing sediment loads in the water.

Taxonomy

<table>
<tr><th>Family</th><th>Image</th><th>Genera</th><th>Description</th></tr>
<tr><td rowspan="2">Zosteraceae</td><td colspan="3">The family Zosteraceae, also known as the seagrass family, includes two genera containing 22 marine species. It is found in temperate and subtropical coastal waters, with the highest diversity around Korea and Japan.

Species subtotal:</td></tr>
<tr><td></td><td>Phyllospadix</td><td>6 species</td></tr>
</table>

Zosteraceae		Zostera	16 species
Hydrocharitaceae	The family **Hydrocharitaceae**, also known as **tape-grasses**, include Canadian waterweed and frogbit. The family includes both fresh and marine aquatics, although of the seventeen species currently recognised only three are marine. They are found throughout the world in a wide variety of habitats, but are primarily tropical. Species subtotal:		
		Enhalus	1 species
		Halophila	19 species
		Thalassia	2 species
Posidoniaceae	The family **Posidoniaceae** contains a single genus with two to nine marine species found in the seas of the Mediterranean and around the south coast of Australia. Species subtotal: 2 to 9		
		Posidonia	2 to 9 species

<table>
<tr><td rowspan="6">Cymodoceaceae</td><td colspan="3">The family Cymodoceaceae, also known as the manatee-grass family, includes only marine species. Some taxonomists do not recognize this family.

Species subtotal:</td></tr>
<tr><td>
</td><td>Amphibolis</td><td>2 species</td></tr>
<tr><td></td><td>Cymodocea</td><td>4 species</td></tr>
<tr><td></td><td>Halodule</td><td>6 species</td></tr>
<tr><td></td><td>Syringodium</td><td>2 species</td></tr>
<tr><td></td><td>Thalassodendron</td><td>3 species</td></tr>
</table>

Ecosystem Services

Seagrass bed with several echinoids, Grahams Harbour, San Salvador Island, Bahamas

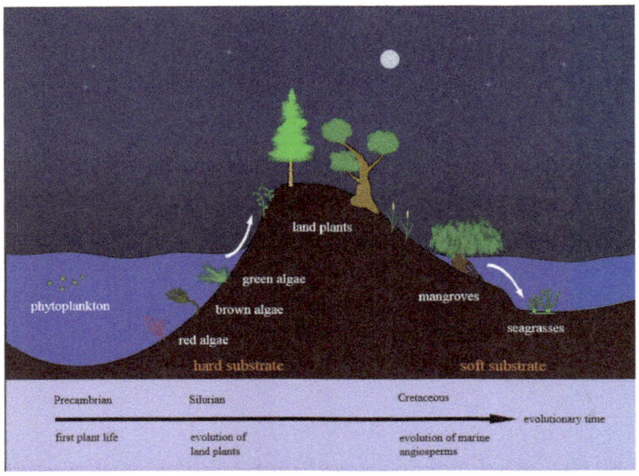

Evolution of seagrasses

Seagrasses are sometimes labeled ecosystem engineers, because they partly create their own habitat: their leaves, by slowing down water currents, increase sedimentation, and their roots and rhizomes stabilize the seabed.

Their importance for associated species is mainly due to provision of shelter (through their three-dimensional structure in the water column) and to their extraordinarily high rate of primary production. As a result, seagrasses provide coastal zones with a number of ecosystem goods and ecosystem services, for instance habitat for commercially and recreationally valued fishery species, fishing grounds, wave protection, oxygen production and protection against coastal erosion. Seagrass meadows account for more than 10% of the ocean's total carbon storage. Per hectare, it holds twice as much carbon dioxide as rain forests. Yearly, seagrasses sequester about 27.4 million tons of CO_2 (Reference Needed). Global warming models suggest, some seagrasses will go extinct – *Posidonia oceanica* is expected to go extinct, or nearly so, by 2050. This would result in CO_2 release.

Uses

Historically, seagrasses were collected as fertilizer for sandy soil. This was an important use in the Ria de Aveiro, Portugal, where the plants collected were known as moliço.

In the early 20th century, in France and, to a lesser extent, the Channel Islands, dried seagrasses were used as a mattress (*paillasse*) filling - such mattresses were in high demand by French forces during World War I. It was also used for bandages and other purposes.

Currently, seagrass is used in furniture, and woven like rattan.

Disturbances and Threats

Natural disturbances, such as grazing, storms, ice-scouring, and desiccation, are an inherent part of seagrass ecosystem dynamics. Seagrasses display an extraordinarily high degree of phenotypic plasticity, adapting rapidly to changing environmental conditions.

Seagrasses are in global decline, with some 30,000 km^2 (12,000 sq mi) lost during recent decades. The main cause is human disturbance, most notably eutrophication, mechanical destruction of habitat, and overfishing. Excessive input of nutrients (nitrogen, phosphorus) is directly toxic to seagrasses, but most importantly, it stimulates the growth of epiphytic and free-floating macro- and micro-algae. This weakens the sunlight, reducing the photosynthesis that nourishes the seagrass and the primary production results.

Decaying seagrass leaves and algae fuels increasing algal blooms, resulting in a positive feedback. This can cause a complete regime shift from seagrass to algal dominance. Accumulating evidence also suggests that overfishing of top predators (large predatory fish) could indirectly increase algal growth by reducing grazing control performed by mesograzers, such as crustaceans and gastropods, through a trophic cascade.

Macro algal blooms cause the decline and eradication of seagrasses throughout areas where nutrient loading or other sources of stimulated algal growth exist. Known as nuisance species, macroalgae grow in filamentous and sheet-like forms and form thick unattached mats over the seagrasse, occurring as epiphytes on seagrass leaves. Eutrophication leads to the forming of a bloom, causing the attenuation of light in the water column, which eventually leads to anoxic conditions for the seagrass and organisms living in/around the plant(s). In addition to the direct blockage of light to the plant, benthic macroalgae have low carbon/nitrogen content, causing their decomposition to stimulate bacterial activity, leading to sediment resuspension, an increase in water turbidity, and the further attenuation of light.

When humans drive motor boats over shallow seagrass areas, sometimes the propeller blade can tear out or cut the seagrass.

The most-used methods to protect and restore seagrass meadows include nutrient and pollution reductions, protection using marine protected areas, and restoration using seagrass transplantation. There is also increasing recognition of the need to increase the resilience of seagrass to the impacts of future environmental change.

Potential Disease Fighting Properties

In February 2017, Cornell University marine biologist and ecologist Dr. Joleah Lamb found that seagrass meadows may be able to remove various disease pathogens from seawater. Dr. Lamb and her research team studied small islands without wastewater treatment facilities in central Indonesia, and found that levels of potentially pathogenic marine bacteria – like *Enterococcus* – that affect humans, fishes, and invertebrates were reduced by 50 percent when seagrass meadows were present compared to paired sites without seagrass.

Thalassia Testudinum

Thalassia testudinum, commonly known as turtlegrass, is a species of marine seagrass. It forms meadows in shallow sandy or muddy locations in the Caribbean Sea and the Gulf of Mexico.

Description

Thalassia testudinum is a perennial grass growing from a long, jointed rhizome that may be buried 25 centimetres (9.8 in) deep in the substrate but is more usually found 5 to 10 cm (2 to 4 in) below the surface. Some nodes are leafless but others bear a tuft of several erect, linear leaf blades. These are up to 30 centimetres (12 in) long and 2 cm (0.8 in) wide and have rounded tips. The flowers grow on short stalks in the axils of the leaves and are greenish-white, sometimes tinged pink, and are followed by seed pods.

Distribution and Habitat

Turtle grass is found growing in meadows in calm shallow waters throughout the Caribbean Sea and the Gulf of Mexico and as far north as Cape Canaveral in Florida. Extensive meadows can be formed on muddy sand, coarse sandy and clayey seabeds, especially those with a calcareous content. This grass favours high-salinity waters with low turbidity such as calm lagoons. It cannot grow in fresh water but some growth is possible at a salinity of 10 parts per thousand. The plant's preferred range is 25 to 38.5 parts per thousand with a temperature range of 20 to 30 °C (68 to 86 °F). It is found from low-tide mark down to depths of 30 metres (98 ft) depending on water clarity. It often grows in meadows with other seagrasses where it is the climax species.

Reproduction

Turtle grass can reproduce both vegetatively and by sexual reproduction. The main propagative method is by increase in length of the rhizomes. This mainly takes place in spring and early summer but can happen at any time of year and results in an increase in the size of the turtle grass bed. It has been found that where plants have been damaged mechanically, such as by the propellers of boats, the cut ends of rhizomes are unable to grow and holes may develop in the turtle grass meadow.

Unusually for the marine environment, turtle grass is a flowering plant. In the spring and early summer, many turtle grass plants produce small flowers at the base of the leaves. Male and female flowers grow on separate plants. Fruits develop in two to four weeks and became detached and float away after about eight weeks.

It has been recently discovered that zooplankton acts as a pollinator for turtle grass. Plankton is drawn to the seagrass's nutritious mucilage—a carbohydrate-rich substance that houses pollen. As the plankton feeds on the mucilage, excess pollen grains stick to their bodies. Plankton then moves from seagrass to seagrass feeding and spreading the pollen.

The seeds are viviparous, and can start new patches of seagrass, but it is thought that the main method of reproduction is vegetative.

Seagrass bed with dense *Thalassia testudinum* and an immature queen conch (*Eustrombus gigas*), Rice Bay, San Salvador Island, Bahamas

Ecology

Turtle grass and other seagrasses form meadows which are important habitats and feeding grounds. Associated seagrass species include *Halophila engelmannii* and *Syringodium filiforme*. Many epiphytes grow on the grasses, and algae, diatoms and bacterial films cover the surface of the leaf blades. The grass is eaten by turtles, herbivorous parrotfish, surgeonfish and sea urchins while the leaf surface films are a food source for many small invertebrates. Decaying turtle grass leaves are responsible for the majority of detritus in meadow areas. This grass is subject to periodic dieback episodes in the Florida Bay area. One such episode in 1987 killed off a large proportion of the plants and the resulting increased sedimentation and greater growth of epiphytes on the remaining plants caused a secondary dieback event. The areas affected have since been reseeded and planted with rhizomes and have recovered. In general, the population of this grass is stable.

Rhizomatous green algae in the genus *Caulerpa* often live among the grasses and many animals make seagrass meadows their home. These include bivalves and other molluscs, polychaete worms, amphipods and juvenile fish which hide among the leaf blades, sea urchins, crabs and caridean shrimps.

Potamogeton Perfoliatus

Potamogeton perfoliatus (claspingleaf pondweed, perfoliate pondweed, redhead grass) is a perennial aquatic plant in the family Potamogetonaceae occurring in both standing and flowing freshwater habitats. It is widely distributed globally, occurring in all continents except South America and Antarctica.

Description

Perfoliate pondweed grows from a robust creeping perennial rhizome, intermittently producing round stems up to 3 m long. The submerged leaves are oval and translucent, with no stalk, 20–115 mm long and 7–42 mm wide, clasping the stem (perfoliate), a flat apex, and 5-12 veins on either side of the midrib. They vary considerably in colour and may be bright green, dark green, yellowish, olive or brownish. There are no floating leaves. The stipules are rather delicate and usually fall off quite soon after the new leaf has unfurled. The insignificant flowers are produced between June and September. Fruits are 4 mm across, olive green in colour.

Perfoliate pondweed growing in a canal. Note the leaves clasping the stem.

Two subspecies have been described. Subspecies *perfoliatus* occurs in the Old World. In North America, populations on the eastern seaboard and the Gulf States are considered to be subsp. *bupleuroides*. Since many pondweeds, including *P. perfoliatus*, vary naturally in response to growing conditions, it is possible that these variants are ecotypes rather than genuine subspecies. The related *P. richardsonii* was formerly considered a subspecies of *P. perfoliatus*, but is now classed as a separate species.

Like most other broadleaved pondweeds, perfoliate pondweed is tetraploid, with 2n=52.

Over much of its range, perfoliate pondweed is unlikely to be confused with any other pondweed species except perhaps *P. praelongus*. In North America it could be confused with *P. richardsonii*, but the latter has stipules that disintegrate to persistent fibres even on the lower part of the stem, whereas those of *P. perfoliatus* disintegrate entirely.

However, perfoliate pondweed regularly hybridises with other *Potamogeton* species including *P. crispus* (*P.* × *cooperi* (Fryer) Fryer), *P. gramineus* (*P.* × *nitens* Weber), *P.lucens* (*P.* × *salicifolius* Wolfg.), *P. alpinus* (*P.* × *prussicus* Hagstr.), *P. berchtoldii* (*P.* × *mysticus* Morong), *P. wrightii* (*P.* × *anguillanus* Koidz.), *P. maackianus* (*P.* × *leptocephalus* Koidz.), *P. alpinus* (*P.* × *prussicus* Hagstr.), *P. nodosus* (*P.* × *assidens* Z. Kaplan, Zalewska-Gałosz et M. Ronikier), *P. richardsonii* (*P.* × *absconditus* Z. Kaplan, Fehrer & Hellq.), *P. epihydrus* (*P.* × *versicolor* Z. Kaplan, Hellq. and Fehrer) and *P. praelongus* (*P.* × *cognatus* Asch. & Graebn.). The first three of these are quite common. A triple hybrid, *P. gramineus* × *lucens* × *perfoliatus* (*P.* × *torssandrii* (Tiselius) Dörfler), is also known. Many of these hybrids are perennial and long-lived, sometimes occurring in the absence of one or even both parents. Therefore, care should be taken with identification of unusual specimens.

Taxonomy

Potamogeton perfoliatus (perfoliate meaning that the leaf encircles the stem) was one of the original species named by Linnaeus in Species Plantarum (1753).

DNA analysis indicates that *P. perfoliatus* is one of the broad-leaved pondweed clade and is probably most closely related to *P. richardsonii*.

Distribution

Potamogeton perfoliatus has a broad global distribution, occurring in all continents except South America and Antarctica. It has been recorded from Asia (Afghanistan, India, Indonesia (Sumatra), Japan, Kazakhstan, Korea, Kyrgyzstan, Mongolia, Pakistan, Russia, Tajikistan, Uzbekistan), Australia, North America (Eastern and SE USA), Europe south to Spain, Central Italy and the Balkans and including Scandinavia, Britain, Ireland and Iceland, North Africa (Algeria, Tunisia, Egypt), the Near and Middle East. It is widespread in lakes and rivers in the British Isles and Ireland, and is common in much of Eurasia and North America, but has a more scattered distribution towards the southern edge of its range.

Ecology and Conservation

P. perfoliatus grows in a wide range of freshwater habitats including lakes, rivers and streams, large ponds, canals and larger drains and ditches. It does not tolerate drying out, and the most robust plants tend to occur in >1 m water depth. Unlike most other broad-leaved pondweeds it is reasonably tolerant of water flow and so can use running waters to a significant extent. It is not especially sensitive to water chemistry provided that the water remains reasonably clear and is not too base-poor, occurring in such diverse habitats as exposed Scottish lochs and lowland fenland rivers. However, at high nutrient concentrations it is vulnerable to shading from phytoplankton and epiphytic algae and filamentous algae. It may also grow in brackish or estuarine habitats, notably in Chesapeake Bay. However, elevated salt concentrations have a negative effect on both growth and flowering.

A *Potamogeton perfoliatus* bed in Llyn Cregennen Isaf, a high quality lake in Wales (UK)

Waterbirds, fish, water beetles and caddis larvae can all be important grazers of perfoliate pondweed, reducing its biomass or even eliminating it locally.

Perfoliate pondweed is less threatened than other broadleaved pondweeds, but it is listed as Vulnerable in Spain and the Netherlands and Endangered in Flanders. There are ongoing efforts to restore populations in Chesapeake Bay, where this was formerly a co-dominant species.

Potamogeton perfoliatus in the River Dordogne, France

Perfoliate pondweed is the commonest of the so-called *Magnopotamion* group of pondweeds. These are a characteristic floristic component of the protected Habitats Directive habitat Type 'Natural eutrophic lakes with *Magnopotamion*'.

Cultivation

Perfoliate pondweed is not widely cultivated, which is perhaps surprising as it is not ecologically demanding and is an attractive plant. In common with other pondweeds of this group it roots poorly from stem cuttings and is best propagated by division of the rhizomes. It is easy to grow in a garden pond or even in a barrel, but unlike some other submerged plants needs to be able to root in a suitable substrate such as sand or aquatic compost. Keeping nutrient levels low helps to prevent smothering by algae and more invasive aquatic plants. Plants being established should be weighted down initially to allow them to root, and benefit from being planted among other aquatic plants that can act as a 'nursery'. Transplants have a high survival rate once established.

P. perfoliatus may also be established from seed, which should be sown on a suitable substrate in shallow water. The seed should be left uncovered, or covered only with a minimal amount of substrate to hold the seed in place and allow root development, as seed buried more than 1 cm deep either does not germinate or the seedlings die. The seed typically germinates within about 10 days.

Although it is possible to grow perfoliate pondweed in fish ponds, it would be vulnerable to fish predation initially, so should be protected during establishment.

Macrophyte

A macrophyte is an aquatic plant that grows in or near water and is either emergent, submergent, or floating, and includes helophytes (a plant that grows in marsh, partly submerged in water, so that it regrows from buds below the water surface). In lakes and rivers macrophytes provide cover for fish and substrate for aquatic invertebrates, produce oxygen, and act as food for some fish and wildlife.

A helophyte (*Iris pseudacorus*)

Fringing stands of tall vegetation by water basins and rivers may include helophytes. Examples include stands of *Equisetum fluviatile*, *Glyceria maxima*, *Hippuris vulgaris*, *Sagittaria*, *Carex*, *Schoenoplectus*, *Sparganium*, *Acorus*, yellow flag (*Iris pseudacorus*), *Typha* and *Phragmites australis*.

Classification of Macrophytes

Emergent Plants

An emergent plant is one which grows in water but which pierces the surface so that it is partially in air. Collectively, such plants are emergent vegetation.

This habit may have developed because the leaves can photosynthesize more efficiently above the shade of cloudy water and competition from submerged plants but often, the main aerial feature is the flower and the related reproductive process. The emergent habit permits pollination by wind or by flying insects.

There are many species of emergent plants, among them, the reed (*Phragmites*), *Cyperus papyrus*, *Typha* species, flowering rush and wild rice species. These may be found growing in fens but usually less well owing to competition from other plants. Some species, such as purple loosestrife, may grow in water as emergent plants but they are capable of flourishing in fens or simply in damp ground.

As Watershed Health Indicators

A decline in a macrophyte community may indicate water quality problems and changes in the ecological status of the water body. Such problems may be the result of excessive turbidity, herbicides, or salinization. Conversely, overly high nutrient levels may create an overabundance of macrophytes, which may in turn interfere with lake processing.

Macrophyte levels are easy to sample, do not require laboratory analysis, and are easily used for calculating simple abundance metrics.

As Potential Sources of Therapeutic Agents

Phytochemical and pharmacological researches suggest that freshwater macrophytes, such as

Centella asiatica, Nelumbo nucifera, Nasturtium officinale, Ipomoea aquatica and *Ludwigia adscendens*, are promising sources of anticancer and antioxidative natural products.

Hot water extracts of the stem and root of *Ludwigia adscendens*, as well as those of the fruit, leaf and stem of *Monochoria hastata* were found to have lipoxygenase inhibitory activity. Hot water extract prepared from the leaf of *Ludwigia adscendens* exhibits alpha-glucosidase inhibitory activity more potent than that of acarbose.

Functions of Macrophytes in Aquatic System

Macrophytes perform many ecosystem functions in aquatic ecosystems and provide services to human society. One of the important functions performed by macrophyte is uptake of dissolve nutrients (N and P) from water . Macrophytes are widely used in constructed wetlands around the world to remove excess N and P from polluted water . Beside direct nutrient uptake, macrophytes indirectly influence nutrient cycling, especially N cycling through influencing the denitrifying bacterial functional groups that are inhabiting on roots and shoots of macrophytes . Macrophytes promote the sedimentation of suspended solids by reducing the current velocities, impede erosion by stabilising soil surfaces . Macrophytes also provide spatial heterogeneity in otherwise unstructured water column. Habitat complexity provided by macrophytes like to increase the richness of taxonomy and density of both fish and invertebrates .

Isoetid

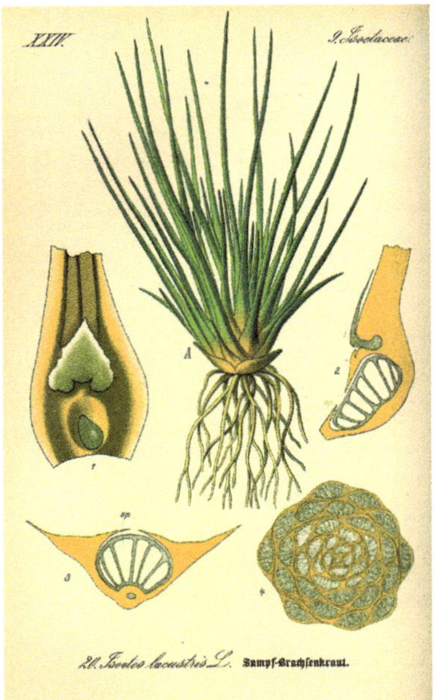

Isoetids are named for their superficial resemblance to, and shared habitat requirements with, the genus *Isoetes*.

Isoetids are aquatic plants or wetland plants named for their superficial similarity to the quillworts, *Isoetes*. They occur in wetlands and on shorelines with low nutrient availability. Owing to their slow growth rates, they are also often found in areas with low rates of sediment deposition. Many have evergreen leaves, and CAM photosynthesis. Often they are exposed during periods of low water. Common examples include *Lobelia dortmanna* and *Eriocaulon septangulare*.

References

- Angiosperm Phylogeny Group (2016). "An update of the Angiosperm Phylogeny Group classification for the orders and families of flowering plants: APG IV". Botanical Journal of the Linnean Society. 181 (1): 1–20. doi:10.1111/boj.12385
- Robert H. Mohlenbrock; Paul Wayne Nelson (1999). "Introduction". Sedges: Carex. Volume 14 of The Illustrated flora of Illinois. Southern Illinois University Press. pp. 3–7. ISBN 978-0-8093-2074-5
- "BSBI List 2007". Botanical Society of Britain and Ireland. Archived from the original (xls) on 2015-01-25. Retrieved 2014-10-17
- Ilkka Kukkonen; Heikki Toivonen (1988). "Taxonomy of wetland carices". Aquatic Botany. 30 (1–2): 5–22. doi:10.1016/0304-3770(88)90003-4
- McGlathery, KJ (2001). "Macroalgal blooms contribute to the decline of seagrass in nutrient-enriched coastal waters" (PDF). Journal of Phycology. 37: 453–456. doi:10.1046/j.1529-8817.2001.037004453.x
- Amjad Almusaed (2010). Biophilic and Bioclimatic Architecture: Analytical Therapy for the Next Generation of Passive Sustainable Architecture. Springer. p. 52. ISBN 978-1-84996-534-7
- "Species Records of Sagittaria". Germplasm Resources Information Network. United States Department of Agriculture. Retrieved 2010-07-13
- Andrew L. Hipp; Paul E. Rothrock; Eric H. Roalson (2009). "The evolution of chromosome arrangements in Carex (Cyperaceae)" (PDF). The Botanical Review. 75 (1): 96–109. doi:10.1007/s12229-008-9022-8
- Horppila, J., Kaitaranta, J., Joensuu, L. & Nurminen, L. (2013) Influence of emergent macrophyte (Phragmites australis) density on water turbulence and erosion of organic-rich sediment. Journal of Hydrodynamics, 25, 288-293
- A. C. Jermy; D. A. Simpson; M. J. Y. Foley; M. S. Porter (2007). "General structure of Cyperaceae". Sedges of the British Isles. BSBI Handbook No. 1 (3rd ed.). Botanical Society of the British Isles. pp. 2–26. ISBN 978-0-901158-35-2
- "Genus: Bacopa Aubl.". Germplasm Resources Information Network. United States Department of Agriculture. 2007-10-05. Retrieved 2011-01-18
- Christenhusz, M. J. M.; Byng, J. W. (2016). "The number of known plants species in the world and its annual increase". Phytotaxa. Magnolia Press. 261 (3): 201–217. doi:10.11646/phytotaxa.261.3.1
- Heinjo Lahring (2003). Water and Wetland Plants of the Prairie Provinces. University of Regina Press. p. 114. ISBN 978-0-88977-162-8
- Short, F. T.; Carruthers, T. J. R.; van Tussenbroek, B.; Zieman, J. (2010). "Thalassia testudinum". IUCN Red List of Threatened Species. Version 2012.2. Retrieved 2012-11-08
- David G. Frodin (2004). "History and concepts of big plant genera". Taxon. 53 (3): 753–776. JSTOR 4135449. doi:10.2307/4135449
- Waycott, Michelle; McMahon, Kathryn; Lavery, Paul (2014). A Guide to Southern Temperate Seagrasses. CSIRO Publishing. ISBN 9781486300150
- Swearingen, Jil M. (7 July 2009). "PCA Alien Plant Working Group - Purple Loosestrife (Lythrum salicaria)". National Park Service. Retrieved 24 September 2011

3

A Comprehensive Study of Aquatic Animals

Aquatic animals live within a water body and breathe the oxygen dissolved in the water. Some of the types of aquatic animals are river dolphin, cavefish, freshwater snail and whale. The topics elaborated in this chapter will help in gaining a better perspective about aquatic animals.

Aquatic Animal

Longfin sculpin (*Jordania zonope*)

An *aquatic animal* is an animal, either vertebrate or invertebrate, which lives in water for most or all of its life. Many insects such as mosquitoes, mayflies, dragonflies and caddisflies have aquatic larvae, with winged adults. Aquatic animals may breathe air or extract oxygen from that dissolved in water through specialised organs called gills, or directly through the skin. Natural environments and the animals that live in them can be categorized as aquatic (water) or terrestrial (land).

The term aquatic can in theory be applied to animals that live in either fresh water (fresh water animals) or salt water (marine animals). However, the adjective marine is most commonly used for animals that live in saltwater, i.e. in oceans, seas, etc.

Aquatic animals (especially freshwater animals) are often of special concern to conservationists because of the fragility of their environments. Aquatic animals are subject to pressure from overfishing, destructive fishing, marine pollution and climate change.

Air Breathing Aquatic Animals

In addition to water breathing animals, e.g., fishes, mollusks etc., the term "aquatic animal" can be applied to air-breathing aquatic or sea mammals such as those in the orders Cetacea (whales) and Sirenia (sea cows), which cannot survive on land, as well as to the pinnipeds (true seals, eared seals, and the walrus). The term "aquatic mammal" is also applied to four-footed mammals like the river otter (*Lontra canadensis*) and beavers (family Castoridae), although these are technically amphibious.

Amphibians, like frogs (the order Anura), while requiring water, are separated into their own environmental classification. The majority of amphibians (class Amphibia) have an aquatic larval stage, like a tadpole, but then live as terrestrial adults, and may return to the water to mate.

Certain fish also evolved to breathe air to survive oxygen-deprived water, such as arapaima (family Osteoglossidae) and walking catfish. Some animals has a hole on their back that blows water out for oxygen.

Cavefish

Phreatichthys andruzzii showing the pale colour and lack of eyes typical of cavefish

Cavefish or cave fish is a generic term for fresh and brackish water fish adapted to life in caves and other underground habitats. Related terms are subterranean fish, troglomorphic fish and hypogean fish.

Being aquatic, they are a part of the troglobite group known as stygofauna. The more than 200 scientifically described species of obligate cavefish are found on all continents, except Antarctica. Although widespread as a group, many cavefish species have very small ranges and are seriously threatened. Cavefish are members of a wide range of families and do not form a monophyletic group. Typical adaptations found in cavefish are reduced eyes and pigmentation.

Adaptations

As typical of cavefish, *Typhleotris madagascariensis* is an opportunistic feeder on various invertebrates

Many aboveground fish may enter caves on occasion, but obligate cavefish (fish that require underground habitats) are extremophiles with a number of unusual adaptations known as troglomorphism. In some species, notably the Mexican tetra, shortfin molly, Oman garra, *Indoreonectes evezardi* and a few catfish, both "normal" aboveground and cavefish forms exist. Many adaptions seen in cavefish are aimed at surviving in a habitat with little food. Living in darkness, pigmentation and eyes are useless, or an actual disadvantage because of their energy requirements, and therefore typically reduced in cavefish. Other examples of adaptations are larger fins for more energy-efficient swimming, and a loss of scales, swim bladder and behaviors such as certain types of display. The loss can be complete or only partial, for example resulting in small (but still existing) eyes. In some cases, "blind" cavefish may still be able to see; juvenile Mexican tetras of the cave form are able to sense light via certain cells in the pineal gland (pineal eye). In the most extreme cases, the lack of light has changed the circadian rhythm (24-hour internal body clock) of the cavefish. In the Mexican tetra of the cave form and in *Phreatichthys andruzzii* the circadian rhythm lasts 30 hours and 47 hours, respectively. This may help them to save energy. Without sight, other senses are used and these may be enhanced. Examples include the lateral line for sensing vibrations, chemoreception (via smell and taste buds), electroreception, and mouth suction to sense nearby obstacles (comparable to echolocation). The level of specialized adaptations in a cavefish is generally considered to be directly correlated to the amount of time it has been restricted to the underground habitat, with species that recently arrived showing few adaptations and species with the largest number of adaptations likely being the ones restricted to the habitat for the longest time.

Cavefish are quite small with most species being less than 10 cm (4 in) long and very few able to surpass 20 cm (8 in). At up to about 40 cm (16 in), the blind cave eel is the longest known cavefish. The very limited food resources in the habitat likely prevents larger cavefish species from existing and also means that cavefish in general are opportunistic feeders, taking whatever is available. In their habitat, cavefish are often the top predators, feeding on smaller cave-living invertebrates, or are detrivores without enemies. Cavefish typically have low metabolic rates and may be able to survive long periods of starvation. A captive *Phreatobius cisternarum* did not feed for a year, but remained in good condition. The cave form of the Mexican tetra can built up unusually large fat reserves by "binge eating" in periods where food is available, which then (together with its low metabolic rate) allows it to survive without food for months, much longer than the aboveground form of the species.

Some fish species that live burried in the bottom of aboveground waters, live deep in the sea or in deep rivers have adaptations similar to cavefish, including reduced eyes and pigmentation.

Habitat

Although many cavefish species are restricted to underground lakes, pools or rivers in actual caves, some are found in aquifers and may only be detected by humans when artificial wells are dug into this layer. Most live in areas with low (essentially static) or moderate water current, but there are also species in places with very strong current, such as the waterfall climbing cave fish. Underground waters are often very stable environments with limited variations in temperature (typically near the annual average of the surrounding region), nutrient levels and other factors. Organic compounds generally only occur in low levels and rely on outside sources, such as contained in water that enters the underground habitat from outside, aboveground animals that find their way into caves (deliberately or by mistake) and guano from bats that roost in caves. Cavefish are primarily restricted to freshwater. A few species, notably the cave-dwelling brotulas, *Luciogobius* gobies, *Milyeringa* sleeper gobies and the blind cave eel, live in anchialine caves and several of these tolerate various salinities.

The Mexican blind brotula and other cave-dwelling brotulas are among the few species that live in anchialine habitas

Range and Diversity

The more than 200 scientifically described obligate cavefish species are found in most continents, but there are strong geographic patterns and the species richness varies. The vast majority of species are found in the tropics or subtropics. Cavefish are strongly linked to regions with karst, which commonly result in underground sinkholes and subterranean rivers.

With more than 120 described species, by far the greatest diversity is in Asia, followed by more than 30 species in South America and about 30 species in North America. In contrast, only 9 species are known from Africa, 5 from Oceania, and 1 from Europe. On a country level, China has the greatest diversity with more than 80 species, followed by Brazil with more than 20 species. India, Mexico, Thailand and the United States of America each have 9–12 species. No other country has more than 4 cavefish species.

The Hoosier cavefish from Indiana in the United States was only described in 2014

Being underground, many places where cavefish may live have not been thoroughly surveyed. New cavefish species are described with some regularity and undescribed species are known. As a consequence, the number of known cavefish species has risen rapidly in recent decades. In the early 1990s only about 50 species were known, in 2010 about 170 species were known, and by 2015 this had surpassed 200 species. It has been estimated that the final number might be around 250 obligate cavefish species. For example, the first cavefish in Europe, a *Barbatula* stone loach, was only discovered in 2015 in Southern Germany. Conversely, their unusual appearance means that some cavefish already attracted attention in ancient times. The oldest known description of an obligate cavefish, involving *Sinocyclocheilus hyalinus*, is almost 500 years old.

Obligate cavefish are known from a wide range of families: Characidae (characids), Balitoridae (hill-stream loaches), Cobitidae (true loaches), Cyprinidae (carps and allies), Nemacheilidae (stone loaches), Amblycipitidae (torrent catfishes), Astroblepidae (naked sucker-mouth catfishes), Callichthyidae (armored catfishes), Clariidae (airbreathing catfishes), Heptapteridae (heptapterid catfishes), Ictaluridae (ictalurid catfishes), Kryptoglanidae (kryptoglanid catfish), Loricariidae (loricariid catfishes), Phreatobiidae (phreatobiid catfishes), Trichomycteridae (pencil catfishes), Sternopygidae (glass knifefishes), Amblyopsidae (U.S. cavefishes), Bythitidae (brotulas), Poeciliidae (live-bearers), Synbranchidae (swamp eels), Cottidae (true sculpins), Eleotridae (sleeper gobies) and Gobiidae (gobies). Many of these families are only very distantly related and do not form a monophyletic group, showing that adaptations to a life in caves has happened numerous times among fish. As such, the descriptive term "cavefish" is an example of folk taxonomy rather than scientific taxonomy. Strictly speaking some Cyprinodontidae (pupfish) are also known from sinkhole caves, famously including the Devils Hole pupfish, but these lack the adaptations (e.g., reduced eyes and pigmentation) typically associated with cavefish. Additionally, species from a few families such as Chaudhuriidae (earthworm eels), Glanapteryginae and Sarcoglanidinae live burried in the bottom of aboveground waters, and can show adaptions similar to traditional underground-living (troglobitic) fish. It has been argued that such species should be recognized as a part of the group of troglobitic fish.

Conservation

The cave form of the Mexican tetra is easily bred in captivity and the only cavefish widely available to aquarists

Although cavefish as a group are found throughout large parts of the world, many cavefish species have tiny ranges (often restricted to a single cave or cave system) and are seriously threatened. For example, the Alabama cavefish is only found in the Key Cave and the entire population has been estimated at less than 100 individuals, while the golden cave catfish only is found in the Aigamas cave in Namibia and has an estimated population of less than 400 individuals. *Caecocypris basimi* from Iraq and the Oaxaca cave sleeper from Mexico may already be extinct, as recent surveys have failed to find them. In some other cases, such as the Brazilian blind characid which went unrecorded by ichthyologists from 1962 to 2004, the apparent "rarity" was likely because of a lack of surveys in its range and habitat, as locals considered it relatively common until the early 1990s (more recently, this species appears to truly have declined significantly). Living in very stable environments, cavefish are likely more vulnerable to changes in the water (for example, temperature or oxygen) than fish of aboveground habitats which naturally experience greater variations. The main threats to cavefish are typically changes in the water level (mainly through water extraction or drought), habitat degradation and pollution, but in some cases introduced species and collection for the aquarium trade also present a threat. Cavefish often show little fear of humans and can sometimes be caught with the bare hands. Most cavefish lack natural predators, although larger cavefish may feed on smaller individuals, and cave-living crayfish, crabs, giant water bugs and spiders have been recorded feeding on a few species of cavefish.

Caves in some parts of the world have been protected, which can safeguard the cavefish. In a few cases such as the Omani blind cave fish (Oman garra), zoos have initiated breeding programs as a safeguard. In contrast to the rarer species, the cave form of the Mexican tetra is easily bred in captivity and widely available to aquarists. This is the most studied cavefish species and likely also the most studied cave organism overall. As of 2006, only six other cavefish species have been bred in captivity, typically by scientists.

River Dolphin

River dolphins are a widely distributed group of fully aquatic mammals that reside exclusively in freshwater or brackish water. They are an informal grouping of dolphins, which is a paraphyletic group within the infraorder Cetacea. The river dolphins comprise the extant families Platanistidae (the Indian dolphins), Iniidae (the Amazonian dolphins), and Pontoporiidae (the brackish dolphins). There are five extant species of river dolphins, and two subspecies. River dolphins, alongside other cetaceans, belong to the clade Cetartiodactyla, with even-toed ungulates, and their closest living relatives the hippopotamuses, having diverged about 40 million years ago.

River dolphins are relatively small compared to other dolphins, having evolved to survive in warm, shallow water and strong river currents. They range in size from the 5-foot (1.5 m) long South Asian river dolphin to the 8-foot (2.4 m) and 220-pound (100 kg) Amazon river dolphin. Several species exhibit sexual dimorphism, in that the males are larger than the females. They have streamlined bodies and two limbs that are modified into flippers. River dolphins use their conical-shaped teeth and long beaks to capture fast-moving prey in murky water. They have well-developed hearing that is adapted for both air and water; they do not really rely on vision since the water they swim in is usually very muddy. These species are well-adapted to living in warm, shallow waters, and, unlike other cetaceans, have little to no blubber.

River dolphins are not very widespread; they are all restricted to certain rivers or deltas. This makes them extremely vulnerable to habitat destruction. River dolphins feed primarily on fish. Male river dolphins typically mate with multiple females every year, but females only mate every two to three years. Calves are typically born in the spring and summer months and females bear all the responsibility for raising them. River dolphins produce a variety of vocalizations, usually in the form of clicks and whistles.

River dolphins are rarely kept in captivity; breeding success has been poor and the animals often die within a few months of capture. As of 2015, there are only three river dolphins in captivity.

Taxonomy and Evolution

Classification

Four families of river dolphins (Iniidae, Pontoporiidae, Lipotidae and Platanistidae) are currently recognized, comprising three superfamilies (Inioidea, Lipotoidea and Platanistoidea). Platanistidae, containing the two subspecies of South Asian river dolphin, is the only accepted family of Platanistoidea. Previously, many taxonomists had assigned all river dolphins to a single family, Platanistidae, and treated the Ganges and Indus River dolphins as separate species. A December 2006 survey found no members of *Lipotes vexillifer* (commonly known as the baiji, or Chinese river dolphin) and declared the species functionally extinct. With their disappearance, one of the recently accepted superfamilies, Lipotoidea, has become extinct.

The current classification of river dolphins is as follows:

Life reconstruction of *Arktocara yakataga*, an Allodelphinidae

- Superfamily Platanistoidea
  - Family Platanistidae
    - Genus *Platanista*
      - South Asian river dolphin, *Platanista gangetica*, with two subspecies
        - Ganges River dolphin (*susu*), *P. g. gangetica*
        - Indus River dolphin (*bhulan*), *P. g. minor*
  - Family Allodelphinidae (Oligocene - Miocene)
  - Family Squalodelphinidae (Oligocene to Miocene)
  - Family Squalodontidae (Oligocene to Miocene)
  - Family Waipatiidae (Oligocene to Miocene)
- Superfamily Inioidea
  - Family Iniidae
    - Genus *Inia*
      - Amazon river dolphin (*boto*), *Inia geoffrensis*
        - *Inia geoffrensis geoffrensis*
        - *Inia geoffrensis humbotiana*

- Araguaian river dolphin, *Inia araguaiaensis*
- Bolivian river dolphin, *Inia boliviensis*
- Genus *Meherrinia* (late Miocene)
- Family Pontoporiidae
- Genus *Auroracetus*
- *Auroracetus bakerae*
- Genus *Pontoporia*
- La Plata dolphin (*Franciscana*), *Pontoporia blainvillei*
- Superfamily Lipotoidea
- Family Lipotidae
- Genus *Lipotes*
- *Baiji* (or Chinese river dolphin), *Lipotes vexillifer* (functionally extinct since December 2006)

In 2012 the Society for Marine Mammalogy began considering the Bolivian (*Inia geoffrensis boliviensis*) and Amazonian (*Inia geoffrensis geoffrensis*) subspecies as full species *Inia boliviensis* and *Inia geoffrensis*, respectively; however, much of the scientific community, including the IUCN, continue to consider the Bolivian population to be a subspecies of *Inia geoffrensis*.

In October 2014, the Society for Marine Mammalogy took *Inia boliviensis* and *Inia araguaiaensis* off their list of aquatic mammal species and subspecies and currently does not recognize these species-level separations.

Evolution

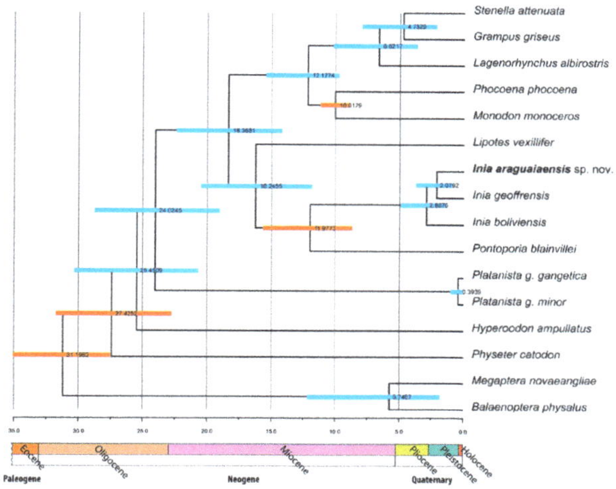

Phylogeny of cetaceans based on cytochrome b gene sequences, showing the distant relationship between *Platanista* and other river dolphins.

River dolphins are members of the infraorder *Cetacea*, which are descendants of land-dwelling mammals of the order Artiodactyla (even-toed ungulates). They are related to the *Indohyus*, an extinct chevrotain-like ungulate, from which they split approximately 48 million years ago. The primitive cetaceans, or archaeocetes, first took to the sea approximately 49 million years ago and became fully aquatic by 5–10 million years later. It is unknown when river dolphins first ventured back into fresh water.

River dolphins are thought to have relictual distributions, that is, their ancestors originally occupied marine habitats, but were then displaced from these habitats by modern dolphin lineages. Many of the morphological similarities and adaptations to freshwater habitats arose due to convergent evolution; thus, a grouping of all river dolphins is paraphyletic. Amazon river dolphins are actually more closely related to oceanic dolphins than to South Asian river dolphins. *Isthminia panamensis* is an extinct genus and species of river dolphin, living 5.8 to 6.1 million years ago. Its fossils were discovered near Piña, Panama.

River dolphin has been considered a taxonomic description, suggesting an evolutionary relationship among the group, although it is now known that they form two distinct clades. 'True' river dolphins are descendants of ancient evolutionary lineages that evolved in freshwater environments.

Some species of cetacean live in rivers and lakes, but are more closely related to oceanic dolphins or porpoises and entered fresh water more recently. Such species are considered facultative freshwater cetaceans as they can use both marine and freshwater environments. These include species such as the Irrawaddy dolphin, *Orcaella brevirostris*, found in the Mekong, Mahakam, the Irrawaddy Rivers, as well as the Yangtze finless porpoise *Neophocaena phocaenoides asiaeorientalis*.

The tucuxi (*Sotalia fluviatilis*) in the Amazon River is another species descended from oceanic dolphins; however, it does not perfectly fit the label of 'facultative' either, as it occurs only in fresh water. The tucuxi was until recently considered conspecific with the Guiana dolphin (*Sotalia guianensis*), which inhabits marine waters. It may also be true for the Irrawaddy dolphin and the finless porpoise that, although the species may be found in both freshwater and marine environments, individual animals found in rivers may not be able to survive in the ocean, and vice versa. The tucuxi is currently classified as an oceanic dolphin (Delphinidae).

The Franciscana (*Pontoporia blainvillei*) has shown a converse evolutionary pattern, and has an ancient evolutionary lineage in freshwater, but inhabits estuarine and coastal waters.

Biology

Anatomy

Indus river dolphin skull

River dolphins have a torpedo shaped body with a flexible neck, limbs modified into flippers, non-existent external ear flaps, a tail fin, and a small bulbous head. River dolphin skulls have small eye orbits, a long snout and eyes placed on the sides of the head. River dolphins are rather small, ranging in size from the 5-foot (1.5 m) long South Asian river dolphin to the 8-foot (2.4 m) and 220-pound (100 kg) Amazon river dolphin. They all have female-biased sexual dimorphism, with the females being larger than the males. River dolphins are polygynous, meaning male river dolphins typically mate with multiple females every year, but females only mate every two to three years. Calves are typically born in the spring and summer months and females bear all the responsibility for raising them.

River dolphins have conical teeth, used to catch swift prey such as small river fish. They also have very long snouts, with some measuring 23 inches (58 cm), four times longer than most of their oceanic counterparts. They have a two-chambered stomach that is similar in structure to that of terrestrial carnivores. They have fundic and pyloric chambers. Breathing involves expelling stale air from their blowhole, followed by inhaling fresh air into their lungs. They do not have the iconic spout, as this only forms when the warm air exhaled from the lungs meets cold external air, which does not occur in their tropical habitats.

River dolphins have a relatively thin layer of blubber. Blubber can help with buoyancy, protection from predators (they would have a hard time getting through a thick layer of fat), energy for leaner times, and insulation from harsh climates. The habitats of river dolphins lack these needs.

Locomotion

River dolphins have two flippers and a tail fin. These flippers contain four digits. Although river dolphins do not possess fully developed hind limbs, some possess discrete rudimentary appendages, which may contain feet and digits. River dolphins are slow swimmers in comparison to oceanic dolphins, which can travel at speeds up to 35 miles per hour (56 km/h); the tucuxi can only travel at about 14 miles per hour (23 km/h). Unlike other cetaceans, their neck vertebrae are not fused together, meaning they have greater flexibility than other non-terrestrial aquatic mammals, at the expense of speed. This means they can turn their head without actually moving their entire body. When swimming, river dolphins rely on their tail fins to propel themselves through the water. Flipper movement is continuous. River dolphins swim by moving their tail fins and lower bodies up and down, propelling themselves through vertical movement, while their flippers are mainly used for steering. All species have a dorsal fin.

Senses

Biosonar by cetaceans

The ears of river dolphins have specific adaptations to their aquatic environment. In humans, the middle ear works as an impedance equalizer between the outside air's low impedance and the cochlear fluid's high impedance. In river dolphins, and other cetaceans, there is no great difference between the outer and inner environments. Instead of sound passing through the outer ear to the middle ear, river dolphins receive sound through the throat, from which it passes through a low-impedance fat-filled cavity to the inner ear. The ear is acoustically isolated from the skull by air-filled sinus pockets, which allows for greater directional hearing underwater. Dolphins send out high frequency clicks from an organ known as a melon. This melon consists of fat, and the skull of any such creature containing a melon will have a large depression. This allows river dolphins to produce biosonar for orientation. They are so dependent on echolocation that they can survive even if they are blind. Beyond locating an object, echolocation also provides the animal with an idea on the object's shape and size, though how exactly this works is not yet understood. The small hairs on the rostrum of the Amazon river dolphin are believed to function as a tactile sense, possibly to compensate for their poor eyesight.

River dolphins have very small eyes

River dolphins have very small eyes for their size, and do not have a very good sense of sight. In addition, the eyes are placed on the sides of the head, so the vision consists of two fields, rather than a binocular view like humans have. When river dolphins surface, their lens and cornea correct the nearsightedness that results from the refraction of light. They have both rod and cone cells, meaning they can see in both dim and bright light. Most river dolphins have slightly flattened eyeballs, enlarged pupils (which shrink as they surface to prevent damage), slightly flattened corneas and a tapetum lucidum; these adaptations allow for large amounts of light to pass through the eye and, therefore, a very clear image of the surrounding area. They also have glands on their eyelids and an outer corneal layer that act as protection for the cornea.

Olfactory lobes are absent in river dolphins, suggesting that they have no sense of smell.

River dolphins are not thought to have a sense of taste, as their taste buds are atrophied or missing altogether. However, some dolphins have preferences between different kinds of fish, indicating some sort of attachment to taste.

Interactions with Humans

Threats

Development

Development and agriculture have had devastating impacts on the habitats on river dolphins. The total population of Araguaian river dolphins is estimated to be between 600 and 1,500 individuals, and genetic diversity is limited. The ecology of their habitat has been adversely affected by agricultural, ranching and industrial activities, as well as by the use of dams for hydroelectric power. The inhabited section of the Araguaia River probably extends over about 900 miles (1,400 km) out of a total length of 1,300 miles (2,100 km). The Tocantins river habitat is fragmented by six hydroelectric dams, so the population there is at particular risk. Its probable eventual IUCN status is Vulnerable or worse.

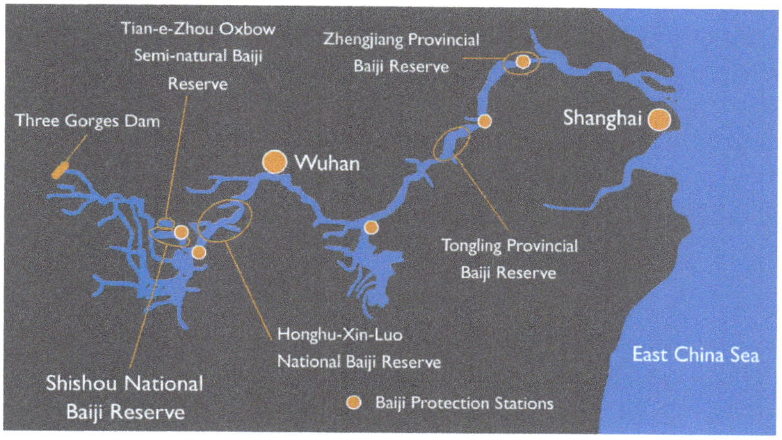

Conservation areas for the baiji along the Yangtze River

Both subspecies of South Asian river dolphins have been very adversely affected by human use of the river systems in the subcontinent. Irrigation has lowered water levels throughout both subspecies' ranges. Poisoning of the water supply from industrial and agricultural chemicals may have also contributed to population decline. Perhaps the most significant issue is the building of more than 50 dams along many rivers, causing the segregation of populations and a narrowed gene pool in which the dolphins can breed. Currently, three subpopulations of Indus river dolphins are considered capable of long-term survival if protected.

As China developed economically, pressure on the baiji river dolphin grew significantly. Industrial and residential waste flowed into the Yangtze. The riverbed was dredged and reinforced with concrete in many locations. Ship traffic multiplied, boats grew in size, and fishermen employed wider and more lethal nets. Noise pollution caused the nearly blind animal to collide with propellers. Stocks of the dolphin's prey declined drastically in the late 20th century, with some fish populations declining to one thousandth of their pre-industrial levels. In the 1950s, the population was estimated at 6,000 animals, but declined rapidly over the subsequent five decades. Only a few hundred were left by 1970. Then the number dropped down to 400 by the 1980s and then to 13 in 1997 when a full-fledged search was conducted. On December 13, 2006, the baiji (*Lipotes vexillifer*) was declared "functionally extinct", after a 45-day search by leading experts in the field failed to find a single specimen. The last verified sighting was in September 2004.

Competition

The region of the Amazon in Brazil has an extension of 3,100,000 sq mi (8,000,000 km²) containing diverse fundamental ecosystems. One of these ecosystems is a floodplain, or a várzea forest, and is home to a large number of fish species which are an essential resource for human consumption. The várzea is also a major source of income through excessive local commercialized fishing. Várzea consist of muddy river waters containing a vast number and diversity of nutrient-rich species. The abundance of distinct fish species lures the Amazon River dolphin into the várzea areas of high water occurrences during the seasonal flooding.

In addition to attracting predators such as the Amazon river dolphin, these high-water occurrences are an ideal location to draw in the local fisheries. Human fishing activities directly compete with the dolphins for the same fish species, the tambaqui (*Colossoma macropomum*) and the pirapitinga (*Piaractus brachypomus*), resulting in deliberate or unintentional catches of the Amazon river dolphin. The local fishermen overfish, and when the Amazon River dolphins remove the commercialized fish from the nets and lines, it damages the equipment and the capture and causes a negative reaction from the local fishermen. The Brazilian Institute of Environment and Renewable Natural Resources prohibit fishermen from killing the Amazon river dolphin, yet they are not compensated for the damage to their equipment and the loss of their catch.

Bycatch

During the process of catching the commercialized fish, the Amazon river dolphins get caught in the nets and exhaust themselves until they die, or the local fishermen deliberately kill the dolphins that become entangled in their nets. The carcasses are discarded, consumed, or used as bait to attract a scavenger catfish, the piracatinga (*Calophysus macropterus*). The use of the Amazon river dolphin carcass as bait for the piracatinga dates back from 2000. The increasing consumption demand by the local inhabitants and Colombia for the piracatinga has created a market for distribution of the Amazon river dolphin carcasses to be used as bait throughout these regions.

For example, of the 15 dolphin carcasses found in the Japurá River in 2010–2011 surveys, 73% of the dolphins were killed for bait, disposed of, or abandoned in entangled gillnets. The data does not fully represent the actual overall number of deaths of the Amazon river dolphins, whether accidental or intentional, because a variety of factors make it extremely complicated to record and medically examine all the carcasses. Scavenger species feed upon them and the complexity of the river currents makes it nearly impossible to locate all the carcasses. More importantly, the local fishermen do not report these deaths out of fear that legal action will be taken against them, as the Amazon river dolphin and other cetaceans are protected under the Brazilian federal law, prohibiting any takes, harassments, and kills of the species.

In Captivity

The only trained Amazon river dolphin in the world at the Acuario de Valencia

A baiji conservation dolphinarium was established at the Institute of Hydrobiology (IHB) in Wuhan in 1992. This was planned as a backup to any other conservation efforts by producing an area completely protected from any threats, and where the baiji could be easily observed. The site includes an indoor and outdoor holding pool, a water filtration system, food storage and preparation facilities, research labs and a small museum. The aim is to also generate income from tourism which can be put towards the baiji plight. The pools are not very large, only kidney shaped tanks with dimensions of 82 feet (25 m) arc 23 feet (7.0 m) width and 11 feet (3.4 m) depth, 33 feet (10 m) diameter, 6.6 feet (2.0 m) deep and 39 feet (12 m) diameter, 11 feet (3.4 m) deep, and are not capable of holding many baijis at one time. Douglas Adams and Mark Carwardine documented their encounters with the endangered animals on their conservation travels for the BBC programme *Last Chance to See*. The book by the same name, published in 1990, included pictures of a captive specimen, a male named Qi Qi that lived in the Wuhan Institute of Hydrobiology dolphinarium from 1980 to July 14, 2002. Discovered by a fisherman in Dongting Lake, he became the sole resident of the Baiji Dolphinarium beside East Lake. A sexually mature female was captured in late 1995, but died after half a year in 1996 when the Shishou Tian-e-Zhou Baiji Semi-natural Reserve, which had contained only finless porpoises since 1990, was flooded.

The Amazon river dolphin has historically been kept in dolphinariums. Today, only three exist in captivity: one in Acuario de Valencia in Venezuela, one in Zoologico de Guistochoca in Peru, and one in Duisburg Zoo in Germany. Several hundred were captured between the 1950s and 1970s, and were distributed in dolphinariums throughout the US, Europe, and Japan. Around 100 went to US dolphinariums, and of that, only 20 survived; the last died in Pittsburgh Zoo in 2002.

In Mythology

Ganga on a river dolphin

Old World

In Hindu mythology, the Ganges River Dolphin is associated with Ganga, the deity of the Ganges river. The dolphin is said to be among the creatures which heralded the goddess' descent from the heavens and her mount, the Makara, is sometimes depicted as a dolphin.

In Chinese mythology, the baiji has many origin stories. For example, near the mouth of the Yangtze, the baiji was a princess that had lost her parents and had lived with her step-father, whom she had longed to get away from. The step-father wanted to trade her since she would have been sold for a great sum of money, but as they were crossing the river to get to the trader, a storm rolled in, and they were drenched. The step-father, enraged, tried to take her, but she plunged herself into the river. Before being drowned in the river, she was transformed into a dolphin, and swam away from her abusive step-father, who also fell in and was transformed into a porpoise.

In another story further upstream the Yangtze, the baiji was the daughter of a general who was deported from the city of Wuhan during a war. During his duty, the daughter ran away. Later, the general met a woman who told him how her father was a general, and when he realized that she was his daughter, he threw himself into the river out of shame. The daughter ran after him and also fell into the river. Before they were drowned, the daughter was transformed into a dolphin, and the general a porpoise.

New World

Amazon river dolphins, known by the natives as the boto or encantados, are very prevalent in the mythology of the native South Americans. They are often characterized in their mythology as wielding superior musical ability, their seductiveness and love of sex, often resulting in illegitimate children, and their attraction to parties. Despite the fact that the Encante are said to come from a utopia full of wealth and without pain or death, the encantados crave the pleasures and hardships of human societies.

Transformation into human form is said to be rare, and usually occurs at night. The encantado will often be seen running from a festa, despite protests from the others for it to stay, and can be seen by pursuers as it hurries to the river and reverts to dolphin form. When it is under human form, it wears a hat to hide its blowhole, which does not disappear with the shapeshift.

Besides the ability to shapeshift into human form, encantados frequently wield other magical abilities, such as the power to control storms, enchant humans into doing their will, transform humans into encantados themselves, and inflict illness, insanity, and even death. Shamans often intervene in these situations.

Kidnapping is also a common theme in such folklore. Encantados are said to be fond of abducting humans with whom they fall in love, children born of their illicit love affairs, or just about anyone near the river who can keep them company, and taking them back to the Encante. The fear of this is so great among people who live in the Amazon river area that both children and adults are terrified of going near the water between dusk and dawn, or entering water-bodies alone. Some who supposedly have encountered encantados while out in their canoes have been said to have gone insane, although in fact, the creatures seem to have done little more than follow their boats and nudge them from time to time.

Amazon River Dolphin

The Amazon river dolphin (*Inia geoffrensis*), also known as the boto, bufeo or pink river dolphin, is a species of toothed whale classified in the family Iniidae. Three subspecies are currently recognized: *I. g. geoffrensis* (Amazon river dolphin), *I. g. boliviensis* (Bolivian river dolphin) and *I. g. humboldtiana* (Orinoco river dolphin). The three subspecies are distributed in the Amazon basin, the upper Madeira River in Bolivia, and the Orinoco basin, respectively.

The Amazon river dolphin is the largest species of river dolphin, with adult males reaching 185 kilograms (408 lb) in weight, and 2.5 metres (8.2 ft) in length. Adults acquire a pink color, more prominent in males, giving it its nickname "pink river dolphin". Sexual dimorphism is very evident, with males measuring 16% longer and weighing 55% more than females. Like other toothed whales, they have a melon, an organ that is used for bio sonar. The dorsal fin, although short in height, is regarded as long, and the pectoral fins are also large. The fin size, unfused vertebrae, and its relative size allow for improved manoeuvrability when navigating flooded forests and capturing prey.

They have one of the widest ranging diets among toothed whales, and feed on up to 53 different species of fish, such as croakers, catfish, tetras and piranhas. They also consume other animals such as river turtles and freshwater crabs.

In 2008, this species was ranked by the International Union for Conservation of Nature (IUCN) as a Data Deficient, due to the uncertainty regarding their population trends and the impact of threats. While hunting is a major threat, in recent decades, greater impacts on population are due to the loss of habitat and inadvertent entanglement in fishing lines.

It is the only species of river dolphin kept in captivity, mainly in the United States, Venezuela and Europe; however, it is difficult to train and a high mortality is seen in captive species.

Taxonomy

An Amazon river dolphin as depicted in *Brehms Tierleben*, 1860's

The species *Inia geoffrensis* was described by Henri Marie Ducrotay de Blainville in 1817. Originally, the Amazon river dolphin belonged to the superfamily Platanistoidea, which constituted all river dolphins, making them a paraphyletic group. Today, however, the Amazon river dolphin has been reclassified into the superfamily Inioidea. There is no consensus on when and how they penetrated the Amazon basin; they may have done so during the Miocene from the Pacific Ocean, before the formation of the Andes, or from the Atlantic Ocean.

Three subspecies are recognized: *I. g. geoffrensis* (Araguaian river dolphin), *I. g. boliviensis* (Bolivian river dolphin) and *I. g. humboldtiana* (Orinoco river dolphin). However, based on skull morphology in 1994, it was proposed that *I. g. boliviensis* was a different species. In 2002, following the analysis of mitochondrial DNA specimens from the Orinoco basin, the Putumayo River (tributary of the Amazon) and the Tijamuchy and Ipurupuru rivers, geneticists proposed that genus *Inia* be divided into at least two evolutionary lineages: one restricted to the river basins of Bolivia and the other widely distributed in the Orinoco and Amazon. A 2014 study supports the notion of the Bolivian group as a distinct species and identifies a third species in the Araguaia-Tocantins basin.

Subspecies

The Bolivian river dolphin is a subspecies of *Inia geoffrensis*

Inia geoffrensis geoffrensis inhabits most of the Amazon River, including rivers Tocantins, Araguaia , low Xingu and Tapajos, the Madeira to the rapids of Porto Velho, and rivers Purus, Yurua, Ica, Caqueta, Branco, and the Rio Negro through the channel of Casiquiare to San Fernando de Atabapo in the Orinoco river, including its tributary: the Guaviare. *Inia humboldtiana geoffrensis* are located in the Orinoco River basin, including the Apure and Meta rivers. This subspecies is restricted, at least during the dry season, to the waterfalls of Rio Negro rapids in the Orinoco between Samariapo and Puerto Ayacucho, and the Casiquiare canal.

Inia geoffrensis boliviensis has populations in the upper reaches of the Madeira River, upstream of the rapids of Teotonio, in Bolivia. It is confined to the Mamore River and its main tributary, the Iténez.

Biology and Ecology

Description

Male Amazon river dolphins are either solid pink or mottled grey/pink

The Amazon river dolphin is the largest river dolphin. Adult males reach a maximum length and weight of 2.55 metres (8.4 ft) (average 2.32 metres (7.6 ft)) and 185 kilograms (408 lb) (average 154 kilograms (340 lb)), while females reach a length and weight of 2.15 metres (7.1 ft) (mean 2 metres (6.6 ft)) and 150 kilograms (330 lb) (average 100 kilograms (220 lb)). It has very evident sexual dimorphism, with males measuring and weighing between 16% and 55% more than females, making it unique among cetaceans, where females are generally larger than males.

The texture of the body is robust and strong but flexible. Unlike oceanic dolphins; the cervical vertebrae are not fused, allowing the head to turn 90 degrees. The flukes are broad and triangular, and the dorsal fin, which is keel-shaped, is short in height but very long, extending from the middle of the body to the caudal region. The pectoral fins are large and paddle-shaped. The length of its fins allows the animal to perform a circular movement, allowing for exceptional maneuverability to swim through the flooded forest but decreasing its speed.

The body color varies with age. Newborns and the young have a dark grey tint, which in adolescence transforms into light grey, and in adults turns pink as a result of repeated abrasion of the skin surface. Males tend to be pinker than females due to more frequent trauma from intra-species aggression. The color of adults varies between solid and mottled pink and in some adults the dorsal surface is darker. It is believed that the difference in color depends on the temperature, water transparency, and geographical location. There is one albino on record, kept in an aquarium in Germany.

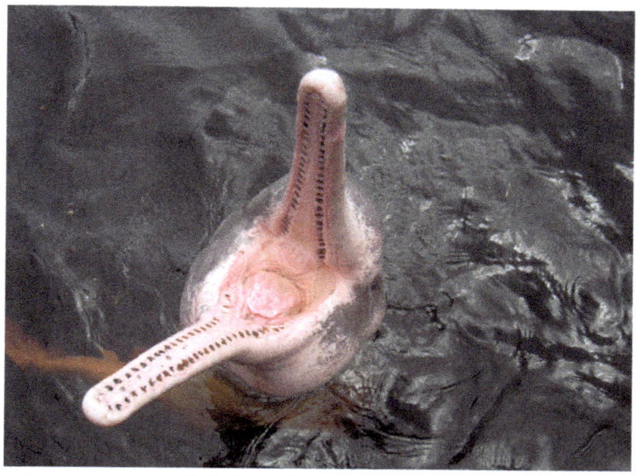

Amazon river dolphins have a heterodont dentition

The skull of the species is slightly asymmetrical compared to the other toothed whales. It has a long, thin snout, with 25 to 28 pairs of long and slender teeth to each side of both jaws. Dentition is heterodont, meaning that the teeth differ in shape and length, with differing functions for both grabbing and crushing prey. Anterior teeth are conical and later have ridges on the inside of the crown. Despite small eyes, the species seems to have good eyesight in and out of the water. It has a melon on the head, the shape of which can be modified by muscular control when used for biosonar. Breathing takes place every 30 to 110 seconds.

Longevity

Life expectancy of the Amazon river dolphin in the wild is unknown, but in captivity, the longevity of healthy individuals has been recorded at between ten and thirty years. However, the average longevity in captive animals is only 33 months. An individual named Apure at Duisburg Zoo, Germany, is estimated to have lived more than forty years, being captured from the wild at age nine.

Apure the dolphin lived for forty years at the Duisburg Zoo

Behavior

The Amazon river dolphin are commonly seen singly or in twos, but may also occur in pods that rarely contain more than eight individuals. Pods as large as 37 individuals have been seen in the Amazon, but average is three. In the Orinoco, the largest observed groups number 30, but average is just above five. Typically, social bonds occur between mother and child, but may also been seen in heterogeneous groups or bachelor groups. The largest congregations are seen in areas with abundant food, and at the mouths of rivers. There is significant segregation during the rainy season, with males occupying the river channels, while females and their offspring are located in flooded areas. However, in the dry season, there is no such separation. Due to the high level of prey fish, larger group-sizes are seen in large sections that are directly influenced by whitewater (such as main rivers and lakes, especially during low water season) than in smaller sections influenced by blackwater (such as channels and smaller tributaries). In their freshwater habitat they are apex predators and gatherings depend more on food sources and habitat availability than in oceanic dolphins where protection from larger predators is necessary.

Captive studies have shown that the Amazon river dolphin is less shy than the bottlenose dolphin, but also less sociable. It is very curious and has a remarkable lack of fear of foreign objects. However, dolphins in captivity may not show the same behavior that they do in their natural environment, where they have been reported to hold the oars of the fishermen, rub against the boat, pluck underwater plants, and play with sticks, logs, clay, turtles, snakes, and fish.

They are slow swimmers; they commonly travel at speeds of 1.5 to 3.2 kilometres per hour (0.93 to 1.99 mph) but have been recorded to swim at speeds up to 14 to 22 kilometres per hour (8.7 to 13.7 mph). When they surface, the tips of the snout, melon and dorsal fins appear simultaneously, the tail rarely showing before diving. They can also shake their fins, and pull their tail fin and head above the water to observe the environment. They occasionally jump out of the water, sometimes as high as a meter (3.14 ft). They are harder to train than most other species of dolphin.

Courtship

Adult males have been observed carrying objects in their mouths such as branches or other floating vegetation, or balls of hardened clay. The males appear to carry these objects as a socio-sexual display which is part of their mating system. The behavior is "triggered by an unusually large number of adult males and/or adult females in a group, or perhaps it attracts such into the group. A plausible explanation of the results is that object carrying is aimed at females and is stimulated by the number of females in the group, while aggression is aimed at other adult males and is stimulated by object carrying in the group." Before determining that the species had an evident sexual dimorphism, it was postulated that the river dolphins were monogamous. Later, it was shown that males were larger than females and are documented wielding an aggressive sexual behavior in the wild and in captivity. Males often have a significant degree of damage in the dorsal, caudal, and pectoral fins, as well as the blowhole, due to bites and abrasions. They also commonly have numerous secondary teeth-raking scars. This suggests fierce competition for access to females, with a polygynous mating system, though polyandry and promiscuity cannot be excluded.

In captivity, courtship and mating foreplay have been documented. The male takes the initiative by nibbling the fins of the female, but reacts aggressively if the female is not receptive. A high frequency of copulations in a couple was observed; they used three different positions: contacting the womb at right angles, lying head to head, or head to tail.

Reproduction

Breeding is seasonal, and births occur between May and June. The period of birthing coincides with the flood season, and this may provide an advantage because the females and their offspring remain in flooded areas longer than males. As the water level begins to decrease, the density of food sources in flooded areas increases due to loss of space, providing enough energy for infants to meet the high demands required for growth. Gestation is estimated to be around eleven months and captive births take 4 to 5 hours. At birth, calves are 80 centimetres (31 in) long and in captivity have registered a growth of 0.21 metres (0.69 ft) per year. Lactation takes about a year. The interval between births is estimated between 15 and 36 months, and the young dolphins are thought to become independent within two to three years.

The relatively long duration of breastfeeding and parenting suggests a strong mother-child bond. Most couples observed in their natural environment consist of a female and her calf. This suggests that long periods of parental care contribute to learning and development of the young.

Diet

The diet of the Amazon river dolphin is the most diverse of the toothed whales. It consists of at least 53 different species of fish, grouped in 19 families. The prey size is between 5 and 80 centimetres (2.0 and 31.5 in), with an average of 20 centimetres (7.9 in). The most frequently consumed fish belong to the families Sciaenidae (croakers), Cichlidae, and Characidae (tetras and piranhas). The dolphin's dentition allows it to access shells of river turtles and freshwater crabs. The diet is more diverse during the wet season, when fish are spread in flooded areas outside riverbeds, thus becoming more difficult to catch. The diet becomes more selective during the dry season when prey density is greater.

Amazon river dolphin feeding

Usually, these dolphins are active and feeding throughout the day and night. However, they are predominantly crepuscular. They consume about 5.5% of their body weight per day. They sometimes take advantage of the disturbances made by boats to catch disoriented prey. Sometimes, they associate with the distantly-related tucuxi (*Sotalia fluviatilis*), and giant otters (*Pteronura brasiliensis*) to hunt in a coordinated manner, by gathering and attacking fish stocks at the same time. Apparently, there is little competition for food between these species, as each prefers different prey. It has also been observed that captive dolphins share food.

Echolocation

Amazonian rivers are often very murky, and the Amazon river dolphin is therefore likely to depend much more on its sense of echolocation than vision when navigating and finding prey. However, echolocation in shallow waters and flooded forests may result in many echoes to keep track of. For each click produced a multitude of echoes are likely to return to the echolocating animal almost on top of each other which makes object discrimination difficult. This may be why the Amazon river dolphin produces less powerful clicks compared to other similar sized toothed whales. By sending out clicks of lower amplitude only nearby objects will cast back detectable echoes, and hence fewer echoes need to be sorted out, but the cost is a reduced biosonar range. Toothed whales generally do not produce a new echolocation click until all relevant echoes from the previous click have been received, so if detectable echoes are only reflected back from nearby objects, the echoes will quickly return, and the Amazon river dolphin is then able to click at a high rate. This in turn allow these animals to have a high acoustic update rate on their surroundings which may aid prey tracking when echolocating in shallow rivers and flooded forests with plenty of hiding places for the prey.

Communication

Like other dolphins, river dolphins use whistling tones to communicate. The issuance of these sounds is related to the time they return to the surface before diving, suggesting a link to food. Acoustic analysis revealed that the vocalisations are different in structure from the typical whistles of other species of dolphins.

Distribution and Population

Amazon river dolphins are the most widespread river dolphins. They are present in six countries in South America: Bolivia, Brazil, Colombia, Ecuador, Peru, and Venezuela, in an area covering about

7,000,000 square kilometres (2,700,000 sq mi). The boundaries are set by waterfalls, such as the Xingu and Tapajós rivers in Brazil, as well as very shallow water. A series of rapids and waterfalls in the Madeira River have isolated one population, recognized as the subspecies *I. g. boliviensis*, in the southern part of the Amazon basin in Bolivia.

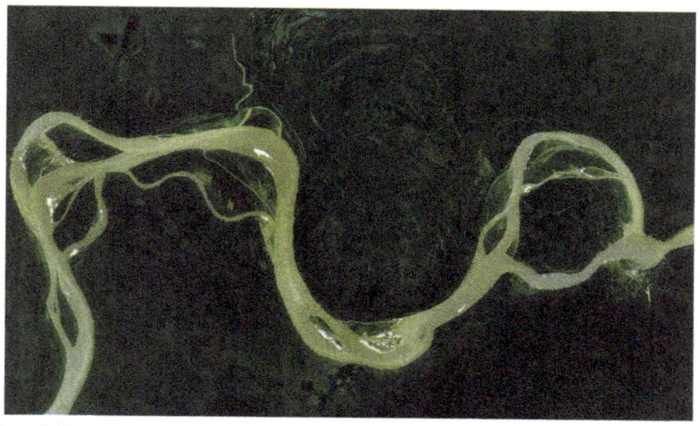

The main branch of the Amazon River near Fonte Boa, Brazil, with multiple floodplains, lakes and smaller channels. The Amazon river dolphin is observed here throughout the year

They are also distributed in the basin of the Orinoco River, except the Caroni River and the upper Caura River in Venezuela. The only connection between the Orinoco and the Amazon is through the Casiquiare canal. The distribution of dolphins in the rivers and surrounding areas depends on the time of year; in the dry season they are located in the river beds, but in the rainy season, when the rivers overflow, they disperse to the flooded areas, both the forests and the plains.

Studies to estimate the population are difficult to analyze due to the difference in the methodology used. In a study conducted in the stretch of the Amazon called Solimões River, with a length of 1,200 kilometres (750 mi) between the cities of Manaus and Tabatinga, a total of 332 individuals was sighted ± 55 per inspection. Density was estimated at 0.08- 0.33 animals per square kilometer in the main channels, and 0.49 to 0.93 animals per square kilometer in the branches. In another study, on a stretch of 120 kilometres (75 mi) at the confluence of Colombia, Brazil and Peru, 345 individuals with a density of 4.8 per square km in the tributaries around the islands. 2.7 and 2.0 were observed along the banks. Additionally, another study was conducted in the Amazon at the height of the mouth of the Caqueta River for six days. As a result of the studies conducted, it was found that the density is higher in the riverbanks, 3.7 per km, decreasing towards the center of the river. In studies conducted during the rainy season, the density observed in the flood plain was 18 animals per square km, while on the banks of rivers and lakes ranged from 1.8 to 5.8 individuals per square km. These observations suggest that the Amazon river dolphin is found in higher density than any other cetacean.

Habitat

The Amazon river dolphin is located in most of the area's aquatic habitats, including; river basins, major courses of rivers, canals, river tributaries, lakes, and at the ends of rapids and waterfalls. Cyclical changes in the water levels of rivers take place throughout the year. During the dry season, dolphins occupy the main river channels, and during the rainy season, they can move easily to smaller tributaries, to the forest, and to floodplains.

Males and females appear to have selective habitat preferences, with the males returning to the main river channels when water levels are still high, while the females and their offspring remain in the flooded areas as long as possible; probably because it decreases the risk of aggression by males toward the young and predation by other species.

Migration

In the Pacaya-Samiria National Reserve, Peru, photo-identification is used to recognize individuals based on pigmentation patterns, scars and abnormalities in the beak. 72 individuals were recognized, of which 25 were again observed between 1991 and 2000. The intervals between sightings ranged from one day to 7.5 years. The maximum range of motion was 220 kilometres (140 mi), with an average of 60.8 kilometres (37.8 mi). The longest distance in one day was 120 kilometres (75 mi), with an average of 14.5 kilometres (9.0 mi). In a previous study conducted at the center of the Amazon River, a dolphin was observed that moved only a few dozen kilometers from the dry season and wet season. However, three of the reviewed 160 animals were observed over 100 kilometres (62 mi) from where they were first registered.

Interactions with Humans

In Captivity

The Amazon river dolphin has historically been kept in dolphinariums. Today, only three exist in captivity: one in Acuario de Valencia in Venezuela, one in Zoologico de Guistochoca in Peru, and one in Duisburg Zoo in Germany. Several hundred were captured between the 1950s and 1970s, and were distributed in dolphinariums throughout the US, Europe, and Japan. Around 100 went to US dolphinariums, and of that, only 20 survived; the last died at the Pittsburgh Zoo in 2002.

Threats

The region of the Amazon in Brazil has an extension of 5,000,000 km² (1,900,000 sq mi) containing diverse fundamental ecosystems. One of these ecosystems is a floodplain, or a várzea forest, and is home to a large number of fish species which are an essential resource for human consumption. The várzea is also a major source of income through excessive local commercialized fishing. Várzea consists of muddy river waters containing a vast number and diversity of nutrient rich species. The abundance of distinct fish species lures the Amazon River dolphin into the várzea areas of high water occurrences during the seasonal flooding.

In addition to attracting predators such as the Amazon river dolphin, these high-water occurrences are an ideal location to draw in the local fisheries. Human fishing activities directly compete with the dolphins for the same fish species, the tambaqui (*Colossoma macropomum*) and the pirapitinga (*Piaractus brachypomus*), resulting in deliberate or unintentional catches of the Amazon river dolphin. The local fishermen overfish and when the Amazon River dolphins remove the commercial catch from the nets and lines, it causes damages to the equipment and the capture, as well as generating ill will from the local fishermen. The negative reactions of the local fishermen are also attributed to the Brazilian Institute of Environment and Renewable Natural Resources prohibition on killing the Amazon river dolphin, yet not compensating the fishermen for the damage done to their equipment and catch.

During the process of catching the commercialized fish, the Amazon river dolphins get caught in the nets and exhaust themselves until they die, or the local fishermen deliberately kill the entangled dolphins. The carcasses are discarded, consumed, or used as bait to attract a scavenger catfish, the piracatinga (*Calophysus macropterus*). The use of the Amazon river dolphin carcass as bait for the piracatinga dates back to 2000. Increasing demand for the piracatinga has created a market for distribution of the Amazon river dolphin carcasses to be used as bait throughout these regions.

Of the 15 dolphin carcasses found in the Japurá River in 2010–2011 surveys, 73% of the dolphins were killed for bait, disposed of, or abandoned in entangled gillnets. The data do not fully represent the actual overall number of deaths of the Amazon river dolphins, whether accidental or intentional, because a variety of factors make it extremely complicated to record and medically examine all the carcasses. Scavenger species feed upon the carcasses, and the complexity of the river currents make it nearly impossible to locate all of the dead animals. More importantly, the local fishermen do not report these deaths out of fear that a legal course of action will be taken against them, as the Amazon river dolphin and other cetaceans are protected under a Brazilian federal law prohibiting any takes, harassments, and kills of the species.

Conservation

In 2008, the International Whaling Commission (IWC) expressed concern for captured botos for use as bait in the Central Amazon, which is an emerging problem that has spread on a large scale. The species is listed in Appendix II of the Convention on International Trade in Endangered Species Fauna and Flora (CITES), and Appendix II of the Convention on the Conservation of Migratory Species of Wild Animals, because it has an unfavorable conservation status or would benefit significantly from international co-operation organized by tailored agreements.

According to a previous assessment by the Scientific Committee of the International Whaling Commission in 2000, the population of botos appears great and there is little or no evidence of population decline in numbers and range. However, increased human intervention on their habitat is expected to, in the future, be the most likely cause of the decline of its range and population. A series of recommendations were issued to ensure proper follow-up to the species, among which is the implementation and publication of studies on the structure of populations, making a record of the distribution of the species, information about potential threats as the magnitude of fishing operations and location of pipelines.

In September 2012, Bolivian President Evo Morales enacted a law to protect the dolphin and declared it a national treasure.

In 2008, the species was listed on the Red list of endangered species, but in 2011, the IUCN stated it as Data Deficient (DD). The species was previously listed as "vulnerable" but the conservation status changed due to the limited amount of currently available information on threats, ecology, and population trends. In areas where these dolphins have been studied, they appear well extended and relatively abundant. However, these areas represent only a small proportion of the total distribution of the species and are often sites where the animals are protected. Consequently, the information from these areas may not be representative, and may not be valid in the long term.

Increasing pollution and gradual destruction of the Amazon rainforest add to the vulnerability of the species. The biggest threats are deforestation and other human activities that contribute to disrupt and alter their environment. Another source of concern is the difficulty in keeping these animals alive in captivity, due to intra-species aggression and low longevity. Captive breeding is not considered a conservation option for this species.

In Mythology

In traditional Amazon River folklore, at night, an Amazon river dolphin becomes a handsome young man who seduces girls, impregnates them, and then returns to the river in the morning to become a dolphin again. Similarly, the female becomes a beautiful, well - dressed,wealthy - looking and young woman. She goes to the house of a married man, places him under a spell to keep him quiet, and takes him to a thatched hut and visits him every year on the same night she seduced him. On the 7th night of visiting, she changes the man into a female - baby or male, and soon transfers it into his own wife's womb. The mythology is said to be the cycle of a baby. This dolphin shapeshifter is called an *encantado*. The myth has been suggested to have arisen partly because dolphin genitalia bear a resemblance to those of humans. Others believe the myth served (and still serves) as a way of hiding the incestuous relations which are quite common in some small, isolated communities along the river. In the area, tales relate it is bad luck to kill a dolphin. Legend also states that if a person makes eye contact with an Amazon river dolphin, he or she will have lifelong nightmares. Local legends also state that the dolphin is the guardian of the Amazonian manatee, and that if one should wish to find a manatee, one must first make peace with the dolphin.

Associated with these legends is the use of various fetishes, such as dried eyeballs and genitalia. These may or may not be accompanied by the intervention of a shaman. A recent study has shown, despite the claim of the seller and the belief of the buyers, none of these fetishes is derived from the *boto*. They are derived from *Sotalia guianensis*, are most likely harvested along the coast and the Amazon River delta, and then are traded up the Amazon River. In inland cities far from the coast, many, if not most, of the fetishes are derived from domestic animals such as sheep and pigs.

Araguaian River Dolphin

The Araguaian river dolphin or Araguaian boto (*Inia araguaiaensis*) is a South American river dolphin population whose identification as a species distinct from the Amazon river dolphin (*Inia geoffrensis*) was announced in 2014. It is native to the Araguaia–Tocantins basin of Brazil.

As of October 2014, *I. araguaiaensis* is not recognized as a separate species by the Society for Marine Mammalogy. Their Committee on Taxonomy made these comments regarding the decision to not include *I. araguaiaensis* as an independent species: "*Inia araguaiaensis*, was described by Hrbek et al. (2014). However, this study only examined samples from two extremes of the distribution of *Inia* so is it unclear if the molecular differences observed represented real species-level separation or were due to sampling from two locations separated by a large distance. Diagnostic osteological differences were also reported. However, because this was based on the examination of very few specimens (only 2 for the new species and only 9 for *I. geoffrensis*), the authors' conclusions are very concerning."

Discovery

The recognition of *I. araguaiaensis* as a separate species was announced on January 22, 2014. It was distinguished from other members of *Inia* on the basis of nuclear microsatellite and mitochondrial DNA data as well as differences in skull morphology (it generally has a wider skull). It also differs from the Amazon and Bolivian river dolphins in the number of teeth per hemimandible (24–28 versus 25–29 and 31–35, respectively). It is the first new river dolphin species to be described since 1918.

Description

Members of the genus are gray to pink in color and have a body length range from 1.53 to 2.6 m (5.0 to 8.5 ft). They have a dorsal ridge rather than a fin. Their neck vertebrae are unfused, allowing them to turn their heads sharply. Like other river dolphins, *I. araguaiaensis* has a prominent forehead and a much longer snout than those of most marine dolphins, as well as smaller eyes than marine dolphins. Because their aquatic environment is often turbid, their vision is not as well developed. River dolphins tend to be less active than marine dolphins. They feed mainly on fish, aided by echolocation.

Taxonomy

This species is most closely related to the Amazon river dolphin (*Inia geoffrensis*), from which it is believed to have split about 2.08 million years (Ma) ago, on the basis of mitochondrial DNA sequence comparisons. The time of divergence corresponds to the time the Amazon and Araguaia-Tocantins river basins became separated, implying vicariant speciation. Major rapids in the lower Tocantins River (into which the Araguaia River flows) are thought to have contributed to isolating the two species, as the Pará River (into which the Tocantins flows) connects with the Amazon River.

The estimated times of divergence with *I. boliviensis*, *P. blainvillei* and *L. vexillifer* are 2.9, 12.0 and 16.2 Ma ago, respectively.

Conservation Issues

An Araguaian river dolphin (*Inia araguaiaensis*) surfacing to breathe in Cantão State Park, Brazil

The total population of the species is estimated to be of the order of 600 to 1500 individuals, and genetic diversity is limited. The ecology of its habitat has been adversely affected by agricultural, ranching and industrial activities, as well as by the use of dams for hydroelectric power. The inhabited section of the Araguaia River probably extends over about 1500 km out of a total length of 2110 km. The Tocantins River habitat is fragmented by six hydroelectric dams, so the population there is at particular risk. The authors of the discovery paper regard its probable eventual IUCN status to be Vulnerable or worse.

The largest number of individuals of the new species is likely to occur in and around Cantão State Park, which contains most of the lakes in the Araguaia basin. However, commercial fishermen around the park have been killing them because they sometimes steal fish from nets. Shooting is common, but around protected areas like Cantão, where the sound of a gun might attract park rangers, some fishermen have taken to putting out poisoned bait for dolphins. The southernmost population of the species is a small group of isolated individuals in the Tocantins river above the Serra da Mesa dam.

Freshwater Snail

Bithynia tentaculata, a small freshwater gastropod in the family Bithyniidae

Freshwater snails are gastropod mollusks which live in freshwater. There are many different families. They are found throughout the world in various habitats, ranging from ephemeral pools to the largest lakes, and from small seeps and springs to major rivers. The great majority of freshwater gastropods have a shell, with very few exceptions. Some groups of snails that live in freshwater respire using gills, whereas other groups need to reach the surface to breathe air. Most feed on algae, but many are detritivors and some are filter feeders.

Pomacea insularum, an apple snail

According to a 2008 review of the taxonomy, there are about 4,000 species of freshwater gastropods (3,795-3,972).

Planorbella trivolvis, an air-breathing ramshorn snail

At least 33–38 independent lineages of gastropods have successfully colonized freshwater environments. It is not possible to quantify the exact number of these lineages yet, because they have yet to be clarified within the Cerithioidea. From six to eight of these independent lineages occur in North America.

Neritimorpha

The Neritimorpha are a group of primitive "prosobranch" gilled snails which have a shelly operculum.

- Neritiliidae, 5 extant freshwater species
- Neritidae, largely confined to the tropics, also the rivers of Europe, family includes the marine "nerites". There are about 110 extant freshwater species.

Family Neritidae, shells of *Theodoxus fluviatilis*.

Family Neritidae, *Neritina natalensis*

Caenogastropoda

The Caenogastropoda are a large group of gilled operculate snails, which are largely marine. In freshwater habitats there are ten major families of caenogastropods, as well as several other families of lesser importance:

Architaenioglossa

- Ampullariidae, an exclusively freshwater family that is largely tropical and includes the large "apple snails" kept in aquaria. 105-170 species.
- Viviparidae, medium to large snails, live-bearing, commonly referred to as "mystery snails". Worldwide except South America, and everywhere confined to fresh waters. 125-150 species.

Family Ampullariidae, *Pomacea bridgesii.*

Family Viviparidae, *Viviparus viviparus.*

Sorbeoconcha

- Melanopsidae, family native to rivers draining to the Mediterranean, also Middle East, and some South Pacific islands. About 25-50 species.
- Pachychilidae - 165-225 species. native to South and Central America. Formerly included with the Pleuroceridae by many authors.
- Paludomidae - about 100 species in south Asia, diverse in African Lakes, and Sri Lanka. Formerly classified with the Pleuroceridae by some authors.
- Pleuroceridae, abundant and diverse in eastern North America, largely high-spired snails of small to large size. About 150 species.

- Semisulcospiridae, - primarily eastern Asia, Japan, also the *Juga* snails of northwestern North America. Formerly included with the Pleuroceridae. About 50 species.
- Thiaridae, high-spired parthenogenic snails of the tropics, includes those referred to as "trumpet snails" in aquaria. About 110 species.

Family Pleuroceridae, *Io fluvialis*.

Family Semisulcospiridae, *Semisulcospira kurodai*.

Family Thiaridae, *Melanoides tuberculata*.

Littorinimorpha

- Littorinidae - 9 species in the genus *Cremnoconchus* are freshwater living in streams and waterfalls. Other species are marine.
- Amnicolidae - about 200 species.
- Assimineidae - about 20 freshwater species, other are marine

- Bithyniidae, small snails, native to Eastern Hemisphere. About 130 species.
- Cochliopidae - 246 species.
- Helicostoidae, the only species *Helicostoa sinensis* lives in China.
- Hydrobiidae, small to very small snails found worldwide. About 1250 freshwater species other are marine.

Clea helena, family Buccinidae.

- Lithoglyphidae - about 100 species.
- Moitessieriidae - 55 species.
- Pomatiopsidae, small amphibious snails scattered worldwide, most diverse in eastern and Southeast Asia. About 170 species.
- Stenothyridae - about 60 freshwater species, others are marine.

Neogastropoda

- Buccinidae - 8-10 freshwater species in the genus *Clea*, native to Southeast Asia. Other Buccinidae are marine.
- Marginellidae - 2 freshwater species in the genus *Rivomarginella*, native to Southeast Asia. Other Marginellidae are marine.

Heterobranchia

Family Valvatidae, shells of *Valvata sibirica*, scale is in mm

Acochlidium fijiensis is one of very few freshwater gastropods without a shell.

Lower Heterobranchia

- Glacidorbidae - 20 species.
- Valvatidae, small low-spired snails referred to as "valve snails". 71 species.

Acochlidiacea

- Acochlidiidae (including synonym Strubelliidae) - 5 shell-less species: *Acochlidium amboinense*, *Acochlidium bayerfehlmanni*, *Acochlidium fijiensis*, *Palliohedyle sutteri* and *Strubellia paradoxa*
- Tantulidae - there is only one species which is shell-less *Tantulum elegans*.

Pulmonata, Basommatophora

Basommatophorans are pulmonate or air-breathing aquatic snails, characterized by having their eyes located at the base of their tentacles, rather than at the tips, as in the true land snails Stylommatophora. The majority of basommatophorans have shells that are thin, translucent, and relatively colorless, and all five freshwater basommatophoran families lack an operculum.

- Chilinidae, small to medium-sized snails confined to temperate and cold South America. About 15 species.
- Latiidae, small limpet-like snails confined to New Zealand. One or three species.
- Acroloxidae - about 40 species.
- Lymnaeidae, found worldwide, but are most numerous in temperate and northern regions. These are the dextral (right-handed) pond snails. About 100 species.
- Planorbidae, "rams horn" snails, with a worldwide distribution. About 250 species.

- Physidae, left-handed (sinistral) "pouch snails", native to Europe, Asia, North America. About 80 species.

Family Lymnaeidae, *Lymnaea stagnalis*.

Family Physidae, *Physella acuta*.

Family Planorbidae, *Planorbarius corneus*.

As Human Food

Several different freshwater snail species are eaten in Asian cuisine.

Archaeological investigations in Guatemala have revealed that the diet of the Maya of the Classic Period (AD 250-900) included freshwater snails.

A dish of cooked saltwater nerites from the Rajang River, Sarawak, Malaysia	A dish of cooked freshwater snails, ampullariids and viviparids from Poipet, Cambodia	A Bengali dish of stir-fried freshwater snails with onion and garlic paste and other spices, from Kolkata, West Bengal, India

Aquarium Snails

In the developed world, people encounter freshwater snails most commonly in aquaria along with tropical fish. Species available vary in different parts of the world. In the United States, commonly available species include ramshorn snails such as *Planorbella duryi*, apple snails such as *Pomacea bridgesii*, the high-spired thiarid malaysian trumpet snail *Melanoides tuberculata*, and several *Neritina* species.

Parasitology

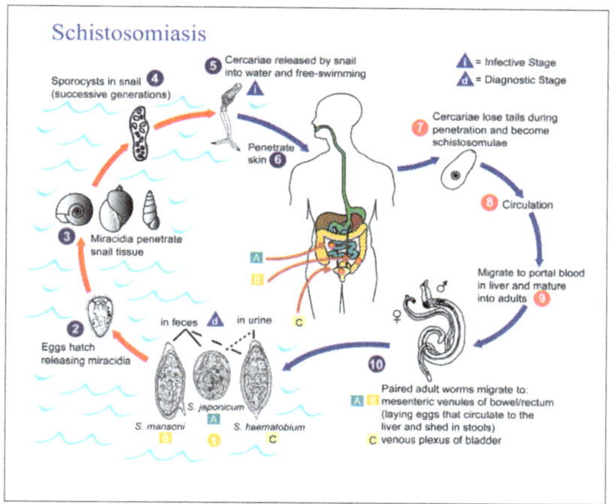

Life cycle of two liver fluke species which have freshwater snails as intermediate hosts

Freshwater snails are widely known to be hosts in the lifecycles of a variety of human and animal parasites, particularly trematodes or "flukes". Some of these relations for prosobranch snails include *Oncomelania* in the family Pomatiopsidae as hosts of *Schistosoma*, and *Bithynia*, *Parafossarulus* and *Amnicola* as hosts of *Opisthorchis*. *Thiara* and *Semisulcospira* may host *Paragonimus*. *Juga plicifera* may host *Nanophyetus salmincola*. Basommatophoran snails are even more widely infected, with many *Biomphalaria* (Planorbidae) serving as hosts for *Schistosoma mansoni*, *Fasciolopsis* and other parasitic groups. The tiny *Bulinus* snails are hosts for *Schistosoma haematobium*. Lymnaeid snails (Lymnaeidae) serve as hosts for *Fasciola* and the cerceriae causing swimmer's itch. The term "neglected tropical diseases" applies to all snail-borne infections, including schistosomiasis, fascioliasis, fasciolopsiasis, paragonimiasis, opisthorchiasis, clonorchiasis, and angiostrongyliasis.

Pomacea Bridgesii

Pomacea bridgesii, common names the spike-topped apple snail or mystery snail, is a South American species of freshwater snail with gills and an operculum, an aquatic gastropod mollusk in the family Ampullariidae.

Subspecies

- *Pomacea bridgesii bridgesii* (Reeve, 1856)
- *Pomacea bridgesii diffusa* (Blume, 1957)

Anatomy

Apple snails possess structurally complex eyes at the tip of a cephalic eyestalk. These snails possess the ability to regenerate the eye completely after amputation through the mid-eyestalk. They are born with both gills and lungs.

Distribution

The native distribution of this snail is Bolivia, Brazil, Paraguay and Peru.

Non-indigenous Distribution

This species is non-indigenous in Hawaii since 1960 (*Pomacea bridgesii diffusa*), southeast Asia since the 1980s, and Florida since the early 1980s (*Pomacea bridgesii diffusa*).

Offspring

Most apple snails lay their eggs above the water line. The eggs take 2–4 weeks to hatch. The snails can produce as many as two-hundred offspring from one egg-laying event. Sometimes not all of the eggs are fertilized so they don't all hatch. When they do hatch, the hatchlings run the risk of getting eaten if they share an aquarium with fish.

Human Relevance

This species is often kept as an aquarium pet, because of its wide range of shell colors, lack of appetite for live plants, and ease of care.

Viviparus Viviparus

Viviparus viviparus is a species of large freshwater snail with a gill and an operculum, an aquatic gastropod mollusk in the family Viviparidae, the river snails.

Description

The height of the shell is 25–35 mm. The width of the shell is 20–26 mm. Males are 2 mm smaller than females of the same age. The shell colour is dark greenish brown or greyish yellow with three reddish brown spiral bands. It is striated but not with a hammer pattern. The shell apex is blunt (more pointed in other *Viviparus* species). The shell has 5.5-6 weakly convex whorls. The last whorl is relatively large compared to that of other *Viviparus* species. The umbilicus is narrow. The animal can lock itself behind a round lid adorned with concentric striations (the operculum) , allowing it to protect itself-for several months if necessary - from dehydration. Once closed the operculum is flush with the opening of the shell.The operculum is attached to the dorsum and to the rear foot of this mollusc, the body of which is wide and T shaped and whose colour varies from spotted grey to greenish to orange.His mouth has a radula and a respiratory siphon opens on the right side of the head, feeding gills located in front of the heart. This trap also allows the animal to filter the water.There are two short tentacles; the eyes are on the external side of each tentacle.

Drawing of the animal and shell of *Viviparus viviparus*; a) head b) tentacles c) eyes d) foot e) operculum

Distribution

This is a European species, which is found in Croatia, Czech Republic (in Bohemia only), Germany, Netherlands, Poland, Great Britain, Ireland and other countries.

Habitat

Viviparus viviparus is largely confined to major, slow-moving, lowland rivers and to lakes and prefers calcareous (base-rich) waters. They are often found in deep water.They are sometimes found in dense clusters (reaching thousands of individuals) on submerged branches and on various man-made objects present under water. More rarely, they are present more scattered in bottom mud, and then are much more discreet. They are also found in canals, artificial ponds, the water behind dams and in reservoirs but usually not in small isolated standing waters. They require high oxygen content.

Feeding

Viviparus viviparus species feeds on plankton and organic microdebris in suspension in the water and picked up through the siphon which allows the animal to breathe while filtering the water.This filter feeding habit makes it popular with owners of ponds or aquariums where they are known to consume filamentous algae, some microalgae, cyanophytes and waste solids and thus help to purify and clarify the water.They may however carry some parasites.

Reproduction

As its latin name suggests, it is a viviparous (oviviparous) snail, a rare phenomenon among snails.The female gives birth to live young, after producing eggs that hatch internally. The naturalist Jan Swammerdam, was the first to recognize the viviparous character of this species to which he gave the name of *Cochlea mirabilis* and *Cochlea vivipara* but he seems not even have understood that there were males and females in this species (most other snails are hermaphroditic).

Then in 1863 Émile Baudelot clearly states that among the paludines (river snails) there are two distinct sexes "The male system extends from the anterior end of the right tentacle to the top of the spire. We may consider it four distinct portions, which are going from top to bottom, the testis, vas deferens, seminal reservoir and the penis.

The male is distinguished by a shorter and round tipped right tentacle, which also serves as a penis during fertilization:the female is usually slightly larger than the male at the same age, and it has two identical tentacles. Sexual maturity is reached after two years, when the snail is about 2 cm long.Each female bears eggs (up to 30 and at all stages of development) with a size of 3 to 7 mm in diameter and up to the full development of the embryo. At the time of their expulsion, the young are about 7 mm and their shell is already marked with the characteristic stripes of the river snails. After producing all its young, the female dies.

In 1879 Mathias Duval made studies of spermatogenesis in *Viviparus viviparus*)

Other life

Principle predators are fish et certain insects (Coleoptera, Hemiptera etc.) *Viviparus viviparus* is an intermediate host of several species of trematodes which finish their life cycles in mammals and birds.

Aquatic Mammal

An Amazon river dolphin (*Inia geoffrensis*), a member of the infraorder Cetacea of the order Cetartiodactyla

Aquatic and semiaquatic mammals are a diverse group of mammals that dwell partly or entirely in bodies of water. They include the various marine mammals who dwell in oceans, as well as various freshwater species, such as the European otter. They are not a taxon and are not unified by any distinct biological grouping, but rather their dependence on and integral relation to aquatic ecosystems. The level of dependence on aquatic life are vastly different among species, with the Amazonian manatee and river dolphins being completely aquatic and fully dependent on aquatic ecosystems; whereas the Baikal seal feeds underwater but rests, molts, and breeds on land; and the capybara and hippopotamus are able to venture in and out of water in search of food.

A Ladoga seal (*Pusa hispida ladogensis*), a member of the clade Pinnipedia of the order Carnivora

Mammal adaptation to an aquatic lifestyle vary considerably between species. River dolphins and manatees are both fully aquatic and therefore are completely tethered to a life in the water. Seals are semiaquatic; they spend the majority of their time in the water, but need to return to land for important activities such as mating, breeding and molting. In contrast, many other aquatic mammals, such as rhinoceroses, capybara, and water shrews, are much less adapted to aquatic living. Likewise, their diet ranges considerably as well, anywhere from aquatic plants and leaves to small fish and crustaceans. They play major roles in maintaining aquatic ecosystems, beavers especially.

Aquatic mammals were the target for commercial industry, leading to a sharp decline in all populations of exploited species, such as beavers. Their pelts, suited for conserving heat, were taken during the fur trade and made into coats and hats. Other aquatic mammals, such as the Indian rhinoceros, were targets for sport hunting and had a sharp population decline in the 1900s. After it was made illegal, many aquatic mammals became subject to poaching. Other than hunting, aquatic mammals can be killed as bycatch from fisheries, where they become entangled in fixed netting and drown or starve. Increased river traffic, most notably in the Yangtze river, causes collisions between fast ocean vessels and aquatic mammals, and damming of rivers may land migratory aquatic mammals in unsuitable areas or destroy habitat upstream. The industrialization of rivers led to the extinction of the Chinese river dolphin, with the last confirmed sighting in 2004.

Taxonomy and Evolution

Aquatic Mammals Vary Greatly in size and Shape		

Eurasian otter (*Lutra lutra*) in Southwold, Suffolk, England	A partially submerged hippopotamus (*Hippopotamus amphibius*) at the Memphis Zoo	A Yapok (*Chironectes minimus*) taxidermy at the Swedish Museum of Natural History

Groups

- Order Sirenia: sirenians
  - Family *Trichechidae*: manatees
    - Amazonian manatee (*Trichechus inunguis*)
    - African manatee (*Trichechus senegalensis*)
    - Dwarf manatee (*Trichechus pygmaeus*) validity questionable
- Order Cetartiodactyla: even-toed ungulates
  - *Suborder Whippomorpha*
    - Family Platanistidae
      - South Asian river dolphin (*Platanista gangetica*) with two subspecies
        - Ganges river dolphin, or susu (*Platanista gangetica gangetica*)
        - Indus river dolphin, or bhulan (*Platanista gangetica minor*)
    - Family Iniidae
      - Amazon river dolphin, or boto (*Inia geoffrensis*)
      - Araguaian river dolphin (*Inia araguaiaensis*)
    - Family Lipotidae
      - Chinese river dolphin, or baiji (*Lipotes vexillifer*) functionally extinct since December 2006
    - Family Pontoporiidae

            - La Plata dolphin, or franciscana (*Pontoporia blainvillei*)
        - Family Hippopotamidae: hippopotamuses
            - Hippopotamus (*Hippopotamus amphibius*)
            - Pygmy hippopotamus (*Choeropsis liberiensis*)
    - *Suborder Ruminantia*
        - Family Cervidae
            - Moose (*Alces alces*)
- Order Carnivora
    - Family Mustelidae
        - *Subfamily Lutrinae*
            - Eurasian otter (*Lutra lutra*)
            - Hairy-nosed otter (*Lutra sumatrana*)
            - Spotted-necked otter (*Hydrictis maculicollis*)
            - Smooth-coated otter (*Lutrogale perspicillata*)
            - North American river otter (*Lontra canadensis*)
            - Southern river otter (*Lontra provocax*)
            - Neotropical river otter (*Lontra longicaudis*)
            - Giant otter (*Pteronura brasiliensis*)
            - African clawless otter (*Aonyx capensis*)
            - Oriental small-clawed otter (*Aonyx cinerea*)
        - *Subfamily Mustelinae*
            - European mink (*Mustela lutreola*)
            - American mink (*Neovison vison*)
    - Family Phocidae
        - Genus *Pusa*
            - Baikal seal (*Pusa sibirica*)
            - Ladoga seal (*Pusa hispida ladogensis*)
            - Saimaa seal (*Pusa hispida saimensis*)

- Order Rodentia: rodents
  - Suborder Hystricomorpha
    - Capybara (*Hydrochoerus hydrochaeris*)
    - Lesser capybara (*Hydrochoerus isthmius*)
    - Coypu (*Myocastor coypus*)
  - Family Castoridae: beavers
    - North American beaver (*Castor canadensis*)
    - Eurasian beaver (*Castor fiber*)
  - Family Cricetidae
    - Muskrat (*Ondatra zibethicus*)
    - European water vole (*Arvicola amphibius*)
- Order Monotremata: monotremes
  - Platypus (Ornithorhynchus anatinus)
- Order Perissodactyla:odd-toed ungulates
  - Family Rhinocerotidae: rhinoceroses
    - Javan rhinoceros (*Rhinoceros sondaicus*)
    - Indian rhinoceros (*Rhinoceros unicornis*)
- Order Afrosoricida
  - Giant otter shrew (*Potamogale velox*)
- Order Soricomorpha
  - Family Soricidae: shrews
    - Malayan water shrew (*Chimarrogale hantu*)
    - Himalayan water shrew (*Chimarrogale himalayica*)
    - Sunda water shrew (*Chimarrogale phaeura*)
    - Japanese water shrew (*Chimarrogale platycephala*)
    - Chinese water shrew (*Chimarrogale styani*)
    - Sumatran water shrew (*Chimarrogale sumatrana*)
    - Elegant water shrew (*Nectogale elegans*)

      - Mediterranean water shrew (*Neomys anomalus*)
      - Eurasian water shrew (*Neomoys fodiens*)
      - Transcaucasian water shrew (*Neomys teres*)
      - Glacier Bay water shrew (*Sorex alaskanus*)
      - American water shrew (*Sorex palustris*)
      - Pacific water shrew, or marsh shrew (*Sorex bendirii*)
    - Family Talpidae (moles and relatives)
      - Russian desman (*Desmana moschata*)
- Order Didelphimorphia: opossums
    - Family Didelphidae: opossums
      - Lutrine opossum (*Lutreolina crassicaudata*)
      - Yapok (*Chironectes minimus*)

Mesozoic

The teeth of *Castorocauda*, a presumably beaver-like mammal, are different in many ways from all other docodonts, presumably due to a difference in diet. Most docodonts had teeth specialized for an omnivorous diet. The teeth of *Castorocauda* suggest that the animal was a piscivore, feeding on fish and small invertebrates. The first two molars had cusps in a straight row, eliminating the grinding function suggesting that they were strictly for gripping and not for chewing. This feature of three cusps in a row is similar to the ancestral condition in mammal relatives (as seen in triconodonts), but is almost certainly a derived character in *Castorocauda*. These first molars were also recurved in a manner designed to hold slippery prey once grasped. These teeth are very similar to the teeth seen in mesonychids, an extinct group of semi-aquatic carnivorous ungulates, and resemble, to a lesser degree, the teeth of seals.

Illustration of *Castorocauda lutrasimilis*, a semi-aquatic mammal from the Jurassic

Another docodontan, the Late Jurassic *Haldanodon*, has been suggested to be a platypus or desman-like swimmer and burrower, being well adapted to dig and swim and occurring in a wetland environment.

The tritylodontid *Kayentatherium* has been suggested to be semi-aquatic. Unlike *Castorocauda* and *Haldanodon*, it was a herbivore, being probably beaver or capybara-like in habits.

Another lineage of Mesozoic mammals, the eutriconodonts, have been suggested to be aquatic animals with mixed results. *Astroconodon* occurred abundantly in freshwater lacustrine deposits and its molars were originally interpreted as being similar to those of piscivorous mammals like cetaceans and pinnipeds; by extension some researchers considered the possibility that all eutriconodonts were aquatic piscivores. However, Zofia Kielan-Jaworowska and other researchers have latter found that the triconodont molars of eutriconodonts were more suited for a carnassial-like shearing action than the piercing and gripping function of piscivorous mammal molars, occluding instead of interlocking, and that *Astroconodon*'s aquatic occurrences may be of little significance when most terrestrial tetrapod fossils are found in lacustrine environments anyway.

However, two other eutriconodonts, *Dyskritodon* and *Ichthyoconodon*, occur in marine deposits with virtually no dental erosion, implying that they died *in situ* and are thus truly aquatic mammals. Nonetheless, *Ichthyoconodon* may not be aquatic, but instead a gliding or even flying mammal. More recently, *Yanoconodon* and *Liaoconodon* have been interpreted as semi-aquatic, bearing a long body and paddle-like limbs.

Life restoration of the eutriconodont *Liaoconodon*.

A metatherian, the stagodontid *Didelphodon*, has been suggested to be aquatic, due to molar and skeleton similarities to sea otters.

Cenozoic

An extinct genus, *Satherium*, is believed to be ancestral to South American river otters, having migrated to the New World during the Pliocene or early Pleistocene. The South American continent houses the *Lontra* genus of otters: the giant otter, the neotropical river otter, the southern river otter, and the marine otter. The smooth-coated otter (*Lutrogale perspicillata*) of Asia may be its closest extant relative; similar behaviour, vocalizations, and skull morphology have been noted.

The most popular theory of the origins of Hippopotamidae suggests that hippos and whales shared a common ancestor that branched off from other artiodactyls around 60 million years ago (mya). This hypothesized ancestral group likely split into two branches around 54 mya. One branch would evolve into cetaceans, possibly beginning about 52 mya, with the protowhale *Pakicetus* and other early whale ancestors collectively known as Archaeoceti, which eventually underwent aquatic adaptation into the completely aquatic cetaceans. The other branch became the anthracotheres, and all branches of the anthracotheres, except that which evolved into Hippopotamidae, became extinct during the Pliocene without leaving any descendants. River dolphins are thought to have relictual distributions, that is, their ancestors originally occupied marine habitats, but were then displaced from these habitats by modern dolphin lineages. Many of the morphological similarities and adaptations to freshwater habitats arose due to convergent evolution; thus, a grouping of all river dolphins is paraphyletic. For example, Amazon river dolphins are actually more closely related to oceanic dolphins than to South Asian river dolphins.

Illustration of *Prorastomus*, a sirenian from the Eocene

Sirenians, along with Proboscidea (elephants), group together with the extinct Desmostylia and likely the extinct Embrithopoda to form the Tethytheria. Tethytheria is thought to have evolved from primitive hoofed mammals ("condylarths") along the shores of the ancient Tethys Ocean. Tethytheria, combined with Hyracoidea (hyraxes), forms a clade called Paenungulata. Paenungulata and Tethytheria (especially the latter) are among the least controversial mammalian clades, with strong support from morphological and molecular interpretations. That is, elephants, hyraxes, and manatees share a common ancestry. The ancestry of Sirenia is distinct from that of Cetacea and Pinnipedia, although they are thought to have evolved an aquatic lifestyle around the same time.

The oldest fossil of the modern platypus dates back to about 100,000 years ago, during the Quaternary period. The extinct monotremes *Teinolophos* and *Steropodon* were once thought to be closely related to the modern platypus, but more recent studies show that platypi are more related to the modern echidnas than to these ancient forms and that at least *Teinolophos* was a rather different mammal lacking several speciations seen in platypi. However, the last common ancestor between platypi and echidnas probably was aquatic, and echidnas thus secondarily became terrestrial. *Monotrematum sudamericanum* is currently the oldest aquatic monotreme known. It has been found in Argentina, indicating monotremes were present in the supercontinent of Gondwana when the continents of South America and Australia were joined via Antarctica, or that monotremes existed along the shorelines of Antarctica in the early Cenozoic.

Marine Mammals

Marine mammals are aquatic mammals that rely on the ocean for their existence. They include animals such as sea lions, whales, dugongs, sea otters and polar bears. Like other aquatic mammals, they do not represent a biological grouping.

The humpback whale is a fully aquatic marine mammal

Marine mammal adaptation to an aquatic lifestyle vary considerably between species. Both cetaceans and sirenians are fully aquatic and therefore are obligate ocean dwellers. Pinnipeds are semiaquatic; they spend the majority of their time in the water, but need to return to land for important activities such as mating, breeding and molting. In contrast, both otters and the polar bear are much less adapted to aquatic living. Likewise, their diet ranges considerably as well; some may eat zooplankton, others may eat small fish, and a few may eat other mammals. While the number of marine mammals is small compared to those found on land, their roles in various ecosystems are large. They, namely sea otters and polar bears, play important roles in maintaining marine ecosystems, especially through regulation of prey populations. Their role in maintaining ecosystems makes them of particular concern considering 23% of marine mammal species are currently threatened.

Marine mammals were first hunted by aboriginal peoples for food and other resources. They were also the target for commercial industry, leading to a sharp decline in all populations of exploited species, such as whales and seals. Commercial hunting lead to the extinction of Steller's sea cow and the Caribbean monk seal. After commercial hunting ended, some species, such as the gray whale and northern elephant seal, have rebounded in numbers, however the northern elephant seal has a genetic bottleneck; conversely, other species, such as the North Atlantic right whale, are critically endangered. Other then hunting, marine mammals, dolphins especially, can be killed as bycatch from fisheries, where they become entangled in fixed netting and drown or starve. Increased ocean traffic causes collisions between fast ocean vessels and large marine mammals. Habitat degradation also threatens marine mammals and their ability to find and catch food. Noise pollution, for example, may adversely affect echolocating mammals, and the ongoing effects of global warming degrades arctic environments.

Adaptations

Mammals evolved on land, so all aquatic and semiaquatic mammals have brought many terrestrial adaptations into the waters. They do not breathe underwater as fish do, so their respiratory systems had to protect the body from the surrounding water; valvular nostrils and an intranarial larynx exclude water while breathing and swallowing. To navigate and detect prey in murky and turbid waters, aquatic mammals have developed a variety of sensory organs: for example, manatees have elongated and highly sensitive whiskers which are used to detect food and other vegetation directly front of them, and toothed whales have evolved echolocation.

Aquatic mammals also display a variety of locomotion styles. Cetaceans excel in streamlined body shape and the up-and-down movements of their flukes make them fast swimmers; the tucuxi, for example, can reach speeds of 14 miles per hour (23 km/h). The considerably slower sirenians can also propel themselves with their fluke, but they can also walk on the bottom with their forelimbs. The earless seals (Phocidae) swim by moving their hind-flippers and lower body from side to side, while their fore-flippers are mainly used for steering. They are clumsy on land, and move on land by lunging, bouncing and wiggling while their fore-flippers keep them balanced; when confronted with predators, they retreat to the water as freshwater phocids have no aquatic predators.

The pygmy hippopotamus has four weight-bearing limbs, and can walk on land like a fully terrestrial mammal.

Some aquatic mammals have retained four weight-bearing limbs (e.g. hippopotamuses, beavers, otters, muskrats) and can walk on land like fully terrestrial mammals. The long and thin legs of a moose limit exposure to and friction from water in contrast to hippopotamuses who keep most of their body submerged and have short and thick legs. The semiaquatic pygmy hippopotamus can walk quickly on a muddy underwater surface thanks to robust muscles and because all toes are weight-bearing. Some aquatic mammals with flippers (e.g. seals) are amphibious and regularly leave the water, sometimes for extended periods, and maneuver on land by undulating their bodies to move on land, similar to the up-and-down body motion used underwater by fully aquatic mammals (e.g. dolphins and manatees).

Beavers, muskrats, otters, capybara have fur, one of the defining mammalian features, that is long, oily, and waterproof in order to trap air to provide insulation. In contrast, other aquatic mammals, such as dolphins, manatees, seals, and hippopotamuses, have lost their fur in favor of a thick and dense epidermis, and a thickened fat layer (blubber) in response to hydrodynamic requirements.

Wading and bottom-feeding animals (e.g. moose and manatee) need to be heavier than water in order to keep contact with the floor or to stay submerged, surface-living animals (e.g. otters) need the opposite, and free-swimming animals living in open waters (e.g. dolphins) need to be neutrally buoyant in order to be able to swim up and down the water column. Typically, thick and dense bone is found in bottom feeders and low bone density is associated with mammals living in deep water.

The shape and function of the eyes in aquatic animals are dependent on water depth and light exposure: limited light exposure results in a retina similar to that of nocturnal terrestrial mammals. Additionally, cetaceans have two areas of high ganglion cell concentration ("best-vision areas"), where other aquatic mammals (e.g. seals, manatees, otters) only have one.

Among non-placental mammals, which cannot give birth to fully developed young, some adjustments have been made for an aquatic lifestyle. The yapok has a backwards-facing pouch which seals off completely when the animal is underwater, while the platypus deposits its young on a burrow on land.

Ecology

Keystone Species

Beaver dams restrict water-flow, creating a pond

Beaver ponds have a profound effect on the surrounding ecosystem. Their first and foremost ecological function is as a reservoir for times of drought, and prevent drying of riverbeds. In the event of a flood, beaver ponds slow down water-flow which reduces erosion on the surrounding soil. Beaver dams hold sediment, which reduces turbidity and thereby improving overall water quality downstream. This supplies other animals with cleaner drinking water and prevents degradation of spawning grounds for fish. However, the slower water speed and lack of shade from trees (that have since been cut down to construct the dam), the overall temperature increases. They also house predatory zooplankton which help break down detritus and control algae populations.

Diet

Beavers are herbivores, and prefer the wood of quaking aspen, cottonwood, willow, alder, birch, maple and cherry trees. They also eat sedges, pondweed, and water lilies. Beavers do not hibernate, but rather they store sticks and logs in a pile in their ponds, eating the underbark. The dams they build flood areas of surrounding forest, giving the beaver safe access to an important food supply, which is the leaves, buds, and inner bark of growing trees. They prefer aspen and poplar, but will also take birch, maple, willow, alder, black cherry, red oak, beech, ash, hornbeam and occasionally pine and spruce. They will also eat cattails, water lilies and other aquatic vegetation, especially in the early spring.

Indian rhinoceros are grazers. Their diets consist almost entirely of grasses, but they also eat leaves, branches of shrubs and trees, fruits, and submerged and floating aquatic plants.They feed in the mornings and evenings. They use their prehensile lips to grasp grass stems, bend the stem down, bite off the top, and then eat the grass. They tackle very tall grasses or saplings by walking over the plant, with legs on both sides and using the weight of their bodies to push the end of the plant down to the level of the mouth.

Manatees make seasonal movements synchronized with the flood regime of the Amazon Basin. They are found in flooded forests and meadows during the flood season when food is abundant, and move to deep lakes during the dry season. The Amazonian manatee has the smallest degree of rostral deflection (25° to 41°) among sirenians, an adaptation to feed closer to the water surface.

A moose's diet often depends on its location, but they seem to prefer the new growths from deciduous trees with a high sugar content, such as white birch, trembling aspen and striped maple, among many others. They also eat many aquatic plants such as lilies and water milfoil. To reach high branches, a moose may bend small saplings down, using its prehensile lip, mouth or body. For larger trees a moose may stand erect and walk upright on its hind legs, allowing it to reach plants 14.0 feet (4.26 m) off the ground. Moose are excellent swimmers and are known to wade into water to eat aquatic plants. Moose are thus attracted to marshes and river banks during warmer months as both provide suitable vegetation to eat and water to bathe in. Moose have been known to dive underwater to reach plants on lake bottoms, and the complex snout may assist the moose in this type of feeding. Moose are the only deer that are capable of feeding underwater.

A moose in Siberia feeding on aquatic plants

Hippopotamuses leave the water at dusk and travel inland, sometimes up to 10 km (6 mi), to graze on short grasses, their main source of food. They spend four to five hours grazing and can consume 68 kg (150 lb) of grass each night. Like almost any herbivore, they consume other plants if presented with them, but their diet consists almost entirely of grass, with only minimal consumption of aquatic plants. The pygmy hippopotamus emerges from the water at dusk to feed. It relies on game trails to travel through dense forest vegetation. It marks trails by vigorously waving its tail while defecating to further spread its feces. The pygmy hippo spends about six hours a day foraging for food, and they do not eat aquatic vegetation to a significant extent and rarely eat grass because it is uncommon in the thick forests they inhabit. The bulk of a pygmy hippo's diet consists of ferns, broad-leaved plants and fruits that have fallen to the forest floor. The wide variety of plants pygmy hippos have been observed eating suggests that they will eat any plants available. This diet is of higher quality than that of the common hippopotamus.

The Amazon river dolphin has the most diverse diet among cetaceans, consisting of at least 53 species of fish. They mainly feed on croakers, cichlids, tetras, and pirahnas, but they may also target freshwater crabs and river turtles. South Asian river dolphins mainly eat fish (such as carp, catfish, and freshwater sharks) and invertebrates, mainly prawns.

Generally, all aquatic desmans, shrews, and voles make quick dives and catch small fish and invertebrates. The giant otter shrew, for example, makes quick dives that last for seconds and grabs small crabs (usually no bigger than 2.8 inches (7 cm) across). The Lutrine opossum is the most carnivorous opossum, usually consuming small birds, rodents, and invertebrates. Water voles mainly eat grass and plants near the water and at times, they will also consume fruits, bulbs, twigs, buds, and roots. However, a population of water voles living in Wiltshire and Lincolnshire, England have started eating frogs' legs and discarding the bodies.

Interactions with Humans

Exploitation

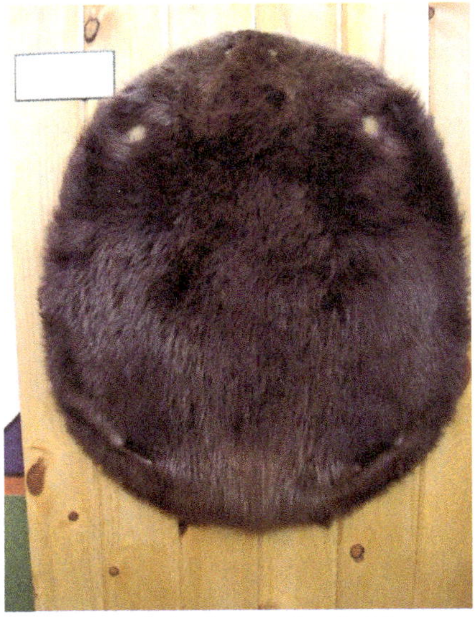

A beaver pelt in the Fur Trade Museum

Fur robes were blankets of sewn-together, native-tanned, beaver pelts. The pelts were called *castor gras* in French and "coat beaver" in English, and were soon recognized by the newly developed felt-hat making industry as particularly useful for felting. Some historians, seeking to explain the term *castor gras*, have assumed that coat beaver was rich in human oils from having been worn so long (much of the top-hair was worn away through usage, exposing the valuable under-wool), and that this is what made it attractive to the hatters. This seems unlikely, since grease interferes with the felting of wool, rather than enhancing it. By the 1580s, beaver "wool" was the major starting material of the French felt-hatters. Hat makers began to use it in England soon after, particularly after Huguenot refugees brought their skills and tastes with them from France.

Sport hunting of the Indian rhinoceros became common in the late 1800s and early 1900s. Indian rhinos were hunted relentlessly and persistently. Reports from the middle of the 19th century claim that some British military officers in Assam individually shot more than 200 rhinos. By 1908, the population in Kaziranga had decreased to around 12 individuals. In the early 1900s, the species had declined to near extinction. Poaching for rhinoceros horn became the single most important reason for the decline of the Indian rhino after conservation measures were put in place from the beginning of the 20th century, when legal hunting ended. From 1980 to

1993, 692 rhinos were poached in India. In India's Laokhowa Wildlife Sanctuary, 41 rhinos were killed in 1983, virtually the entire population of the sanctuary. By the mid-1990s, poaching had rendered the species extinct there. In 1950, Chitwan's forest and grasslands extended over more than 2,600 km^2 (1,000 sq mi) and were home to about 800 rhinos. When poor farmers from the mid-hills moved to the Chitwan Valley in search of arable land, the area was subsequently opened for settlement, and poaching of wildlife became rampant. The Chitwan population has repeatedly been jeopardized by poaching; in 2002 alone, poachers killed 37 animals to saw off and sell their valuable horns.

Otters have been hunted for their pelts since at least the 1700s. There has been a long history of otter pelts being worn around the world. In China it was standard for the royalty to wear robes made from them. People that were financially high in status also wore them. Otters have also been hunted using dogs, specifically the otterhound. In modern times, TRAFFIC, a joint program of the World Wildlife Fund (WWF) and International Union for Conservation of Nature (IUCN), reported that otters are at serious risk in Southeast Asia and have disappeared from parts of their former range. This decline in populations is due to hunting to supply the demand for skins.

Habitat Degradation

The Baykalsk Pulp and Paper Mill was a major producer of industrial waste for Lake Baikal.

One problem at Lake Baikal is the introduction of pollutants into the ecosystem. Pesticides such as DDT and hexachlorocyclohexane, as well as industrial waste, mainly from the Baykalsk Pulp and Paper Mill, are thought to have been the cause of several disease epidemics among Baikal seal populations. The chemicals are speculated to concentrate up the food chain and weaken the Baikal seal's immune system, making them susceptible to diseases such as canine distemper and the plague, which was the cause of a serious Baikal seal epidemic that resulted in the deaths of thousands of animals in 1997 and 1999. Baikal seal pups have higher levels of DDT and PCB than known in any other population of European or Arctic earless seal.

In the 1940s, beavers were brought from Canada to the island of Tierra Del Fuego in southern Chile and Argentina, for commercial fur production. However, the project failed and the beavers, ten pairs, were released into the wild. Having no natural predators in their new environment, they quickly spread throughout the island, and to other islands in the region, reaching a number of 100,000 individuals within just 50 years. They are now considered a serious invasive species in the region, due to their massive destruction of forest trees, and efforts are being made for their eradication.

In some European countries, such as Belgium, France, and the Netherlands, the muskrat is considered an invasive pest, as its burrowing damages the dikes and levees on which these low-lying countries depend for protection from flooding. In those countries, it is trapped, poisoned, and hunted to attempt to keep the population down. Muskrats also eat corn and other farm and garden crops growing near water bodies.

Dead trees as a result of the construction of a beaver dam in Tierra del Fuego

Urban and agricultural development, increased damming, and increased use of hydroelectric power in rivers in countries such as Côte d'Ivoire and Ghana are threats to the African manatee's habitat and life, and thick congestion of boats in waterways may cause them to have a deadly run-in with the vessels. However, even natural occurrences, such as droughts and tidal changes, often strand manatees in an unsuitable habitat. Some are killed accidentally by fishing trawls and nets intended for catching sharks. The Amazonian manatee is at risk from pollution, accidental drowning in commercial fishing nets, and the degradation of vegetation by soil erosion resulting from deforestation. Additionally, the indiscriminate release of mercury in mining activities threatens the entire aquatic ecosystem of the Amazon Basin.

As China developed economically, pressure on the Chinese river dolphin grew significantly. Industrial and residential waste flowed into the Yangtze. The riverbed was dredged and reinforced with concrete in many locations. Ship traffic multiplied, boats grew in size, and fishermen employed wider and more lethal nets. Noise pollution caused the nearly blind animal to collide with propellers. Stocks of the dolphin's prey declined drastically in the late 20th century, with some fish populations declining to one thousandth of their pre-industrial levels. In the 1950s, the population was estimated at 6,000 animals, but declined rapidly over the subsequent five decades. Only a few hundred were left by 1970. Then the number dropped down to 400 by the 1980s and then to 13 in 1997 when a full-fledged search was conducted. On December 13, 2006, the baiji was declared functionally extinct, after a 45-day search by leading experts in the field failed to find a single specimen. The last verified sighting was in 2004.

As Food

Moose are hunted as a game species in many of the countries where they are found. While the flesh has protein levels similar to those of other comparable red meats (e.g. beef, deer and elk), it has a low fat content, and the fat that is present consists of a higher proportion of polyunsaturated fats rather than saturated fats.

"...like tender beef, with perhaps more flavour; sometimes like veal"

—Henry David Thoreau of The Maine Woods describing the taste of moose meat

Cadmium levels are high in moose liver and kidneys, with the result that consumption of these organs from moose more than one year old is prohibited in Finland. Cadmium intake has been found to be elevated amongst all consumers of moose meat, though the meat was found to contribute only slightly to the daily cadmium intake. However the consumption of moose liver or kidneys significantly increased cadmium intake, with the study revealing that heavy consumers of moose organs have a relatively narrow safety margin below the levels which would probably cause adverse health effects.

In the 17th century, based on a question raised by the Bishop of Quebec, the Roman Catholic Church ruled that the beaver was a fish (beaver flesh was a part of the indigenous peoples' diet, prior to the Europeans' arrival) for purposes of dietary law. Therefore, the general prohibition on the consumption of meat on Fridays did not apply to beaver meat. This is similar to the Church's classification of other semi-aquatic rodents, such as the capybara and muskrat.

Marine Mammal

A leopard seal (*Hydrurga leptonyx*), a member of the clade Pinnipedia of the order Carnivora

Marine mammals are aquatic mammals that rely on the ocean and other marine ecosystems for their existence. They include animals such as seals, whales, manatees, sea otters and polar bears. They do not represent a distinct taxon or systematic grouping, but rather have a polyphyletic relation due to convergent evolution. They are also unified by their reliance on the marine environment for feeding.

Marine mammal adaptation to an aquatic lifestyle vary considerably between species. Both cetaceans and sirenians are fully aquatic and therefore are obligate water dwellers. Seals and sea-lions are semiaquatic; they spend the majority of their time in the water, but need to return to land for important activities such as mating, breeding and molting. In contrast, both otters and the polar bear are much less adapted to aquatic living. Their diet varies considerably as well; some may eat zooplankton, others may eat fish, squid, shellfish, sea-grass and a few may eat other mammals. While the number of marine mammals is small compared to those found on land, their roles in various ecosystems are large, especially concerning the maintenance of marine ecosystems, through processes including the regulation of prey populations. This role in maintaining ecosystems makes them of particular concern as 23% of marine mammal species are currently threatened.

Marine mammals were first hunted by aboriginal peoples for food and other resources. Many were also the target for commercial industry, leading to a sharp decline in all populations of exploited species, such as whales and seals. Commercial hunting lead to the extinction of Steller's sea cow and the Caribbean monk seal. After commercial hunting ended, some species, such as the gray whale and northern elephant seal, have rebounded in numbers; conversely, other species, such as the North Atlantic right whale, are critically endangered. Other than hunting, marine mammals can be killed as bycatch from fisheries, where they become entangled in fixed netting and drown or starve. Increased ocean traffic causes collisions between fast ocean vessels and large marine mammals. Habitat degradation also threatens marine mammals and their ability to find and catch food. Noise pollution, for example, may adversely affect echolocating mammals, and the ongoing effects of global warming degrade arctic environments.

Taxonomy

Marine mammals form a diverse group of 129 species that rely on the ocean for their existence. They do not represent a distinct taxon or systematic grouping, but instead have a paraphyletic relationship. They are also unified by their reliance on the marine environment for feeding. The level of dependence on the marine environment for existence varies considerably with species. For example, dolphins and whales are completely dependent on the marine environment for all stages of their life, seals feed in the ocean but breed on land, and polar bears must feed on land. Twenty three percent of marine mammal species are threatened.

Marine Mammals of Varying Sizes and Shapes

A polar bear (*Ursus maritimus*), a member of family Ursidae.

A sea otter (*Enhydra lutris*), a member of family Mustelidae.

A West Indian manatee (*Trichechus manatus*), a member of order Sirenia.

Classification of Extant Species

- Order Cetartiodactyla
    - Suborder Whippomorpha

        - Family Balaenidae (right and bowhead whales), two genera and four species
        - Family Cetotheriidae (pygmy right whale), one species
        - Family Balaenopteridae (rorquals), two genera and eight species
        - Family Eschrichtiidae (gray whale), one species
        - Family Physeteridae (sperm whale), one species
        - Family Kogiidae (pygmy and dwarf sperm whales), one genus and two species
        - Family Monodontidae (narwhal and beluga), two genera and two species
        - Family Ziphiidae (beaked whales), six genera and 21 species
        - Family Delphinidae (oceanic dolphins), 17 genera and 38 species
        - Family Phocoenidae (porpoises), two genera and seven species
- Order Sirenia (sea cows)
    - Suborder Cynodontia
        - Family Trichechidae (manatees), two species
        - Family Dugongidae (dugongs), one species
- Order Carnivora (carnivores)
    - Suborder Caniformia
        - Family Mustelidae, three species
        - Family Ursidae (bears), one species
    - Suborder Pinnipedia (sealions, walruses, seals)
        - Family Otariidae (eared seals), seven genera and 15 species
        - Family Odobenidae (walrus), one species
        - Family Phocidae (earless seals), 14 genera and 18 species

Evolution

Mammals have returned to the ocean in many separate evolutionary lineages, namely: Cetacea, Sirenia, Desmostylia, Pinnipedia, *Kolponomos* (marine bear), *Thalassocnus* (aquatic sloth), *Ursus maritimus* (polar bear), and *Enhydra lutris* (sea otter); the eutriconodonts *Ichthyoconodon* and *Dyskritodon* might have also been marine in habits. Five of these lineages are extinct (Desmostylia; *Kolponomos*; *Thalassocnus*, *Dyskritodon*, *Ichthyoconodon*). Despite the diversity in morphology seen between groups, improving foraging efficiency has been the main driver in the evolution in these lineages. Today, fully aquatic marine mammals belong to one of two orders: Cetartiodactyla or Sirenia. The Cetartiodactyla lineage became aquatic around 50 million years ago (mya), then Sirenia 40 mya, and Pinnipedia around 20 to 25 mya, then sea otters two mya, and most recently the polar bear around 130,000 to 110,000 years ago.

A skeleton of *Thalassocnus* (5–3 mya) from the Muséum national d'histoire naturelle in its presumed swimming pose

Based on molecular and morphological research, the cetaceans genetically and morphologically fall firmly within the Artiodactyla (even-toed ungulates). The term Cetartiodactyla reflects the idea that whales evolved within the ungulates. The term was coined by merging the name for the two orders, Cetacea and Artiodactyla, into a single word. Under this definition, the closest living land relative of the whales and dolphins is thought to be the hippopotamuses. Use of the order Cetartiodactyla, instead of Cetacea with parvorders Odontoceti and Mysticeti, is favored by most evolutionary mammalogists working with molecular data and is supported the IUCN Cetacean Specialist Group and by Taxonomy Committee of the Society for Marine Mammalogy, the largest international association of marine mammal scientists in the world. Some others, including many marine mammalogists and paleontologists, favor retention of order Cetacea with the two suborders in the interest of taxonomic stability.

Pinnipeds split from other caniforms 50 mya during the Eocene. Their evolutionary link to terrestrial mammals was unknown until the 2007 discovery of *Puijila darwini* in early Miocene deposits in Nunavut, Canada. Like a modern otter, *Puijila* had a long tail, short limbs and webbed feet instead of flippers. The lineages of Otariidae (eared seals) and Odobenidae (walrus) split almost 28 mya. Phocids (earless seals) are known to have existed for at least 15 mya, and molecular evidence supports a divergence of the Monachinae (monk seals) and Phocinae lineages 22 mya.

Fossil evidence indicates the sea otter (*Enhydra*) lineage became isolated in the North Pacific approximately two mya, giving rise to the now-extinct *Enhydra macrodonta* and the modern sea otter, *Enhydra lutris*. The sea otter evolved initially in northern Hokkaidō and Russia, and then spread east to the Aleutian Islands, mainland Alaska, and down the North American coast. In comparison to cetaceans, sirenians, and pinnipeds, which entered the water approximately 50, 40, and 20 mya, respectively, the sea otter is a relative newcomer to marine life. In some respects though, the sea otter is more fully adapted to water than pinnipeds, which must haul out on land or ice to give birth.

The first appearance of sirenians in the fossil record was during the early Eocene, and by the late Eocene, sirenians had significantly diversified. Inhabitants of rivers, estuaries, and nearshore marine waters, they were able to spread rapidly. The most primitive sirenian, *Prorastomus*, was found in Jamaica, unlike other marine mammals which originated from the Old World (such as cetaceans). The first known quadrupedal sirenian was *Pezosiren* from the early Eocene. The earliest known sea cows, of the families Prorastomidae and Protosirenidae, were both confined to the Eocene, and were pig-sized, four-legged, amphibious creatures. The first members of Dugongidae appeared by the end of the Eocene. At this point, sea cows were fully aquatic.

Polar bears are thought to have diverged from a population of brown bears, *Ursus arctos*, that became isolated during a period of glaciation in the Pleistocene or from the eastern part of Siberia, (from Kamchatka and the Kolym Peninsula). The oldest known polar bear fossil is a 130,000 to 110,000-year-old jaw bone, found on Prince Charles Foreland in 2004. The mitochondrial DNA (mtDNA) of the polar bear diverged from the brown bear roughly 150,000 years ago. Further, some clades of brown bear, as assessed by their mtDNA, are more closely related to polar bears than to other brown bears, meaning that the polar bear might not be considered a species under some species concepts.

Distribution and Habitat

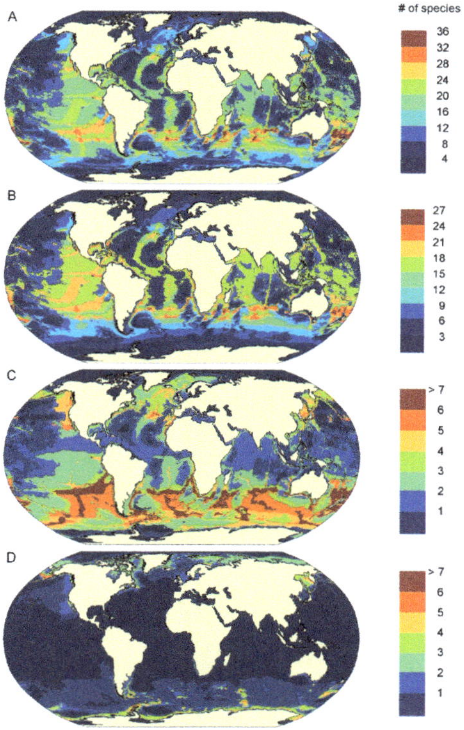

Predicted patterns of marine mammal species richness. A) All species (n=115), B) Odontocetes (n=69), C) Mysticetes (n=14), D) Pinnipeds (n=32). Colors indicate the number of species predicted to occur in each 0.5°x0.5° grid cell from a relative environmental suitability (RES) model, using environmental data from 1990–1999, and assuming a presence threshold of RES>0.6.

Marine mammals are widely distributed throughout the globe, but their distribution is patchy and coincides with the productivity of the oceans. Species richness peaks at around 40° latitude, both north and south. This corresponds to the highest levels of primary production around North and South America, Africa, Asia and Australia. Total species range is highly variable for marine mammal species. On average most marine mammals have ranges which are equivalent or smaller than one-fifth of the Indian Ocean. The variation observed in range size is a result of the different ecological requirements of each species and their ability to cope with a broad range of environmental conditions. The high degree of overlap between marine mammal species richness and areas of human impact on the environment is of concern.

Most marine mammals, such as seals and sea otters, inhabit the coast. Seals, however, also use a number of terrestrial habitats, both continental and island. In temperate and tropical areas, they haul-out on to sandy and pebble beaches, rocky shores, shoals, mud flats, tide pools and in sea caves. Some species also rest on man-made structures, like piers, jetties, buoys and oil platforms. Seals may move further inland and rest in sand dunes or vegetation, and may even climb cliffs. Most cetaceans live in the open ocean, and species like the sperm whale may dive to depths of −1,000 to −2,500 feet (−300 to −760 m) in search of food. Sirenians live in shallow coastal waters, usually living 30 feet (9.1 m) below sea level. However, they have been known to dive to −120 feet (−37 m) to forage deep-water seagrasses. Sea otters live in protected areas, such as rocky shores, kelp forests, and barrier reefs, although they may reside among drift ice or in sandy, muddy, or silty areas.

Many marine mammals seasonally migrate. Annual ice contains areas of water that appear and disappear throughout the year as the weather changes, and seals migrate in response to these changes. In turn, polar bears must follow their prey. In Hudson Bay, James Bay, and some other areas, the ice melts completely each summer (an event often referred to as "ice-floe breakup"), forcing polar bears to go onto land and wait through the months until the next freeze-up. In the Chukchi and Beaufort seas, polar bears retreat each summer to the ice further north that remains frozen year-round. Seals may also migrate to other environmental changes, such as El Niño, and traveling seals may use various features of their environment to reach their destination including geomagnetic fields, water and wind currents, the position of the sun and moon and the taste and temperature of the water. Baleen whales famously migrate very long distances into tropical waters to give birth and raise young, possibly to prevent predation by killer whales. The gray whale has the longest recorded migration of any mammal, with one traveling 14,000 miles (23,000 km) from the Sea of Okhotsk to the Baja Peninsula. During the winter, manatees living at the northern end of their range migrate to warmer waters.

Adaptations

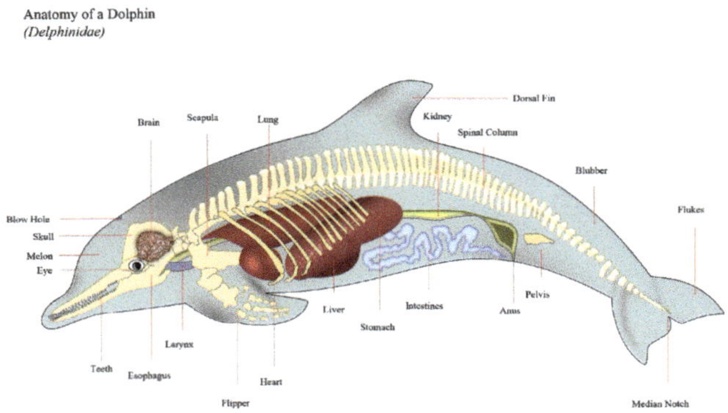

The anatomy of a dolphin showing its skeleton, major organs, and body shape

Marine mammals have a number of physiological and anatomical features to overcome the unique challenges associated with aquatic living. Some of these features are very species specific. Marine mammals have developed a number of features for efficient locomotion such as torpedo shaped bodies to reduce drag; modified limbs for propulsion and steering; tail flukes and dorsal fins for propulsion and balance. Marine mammals are adept at thermoregulation using dense fur or blubber, circulatory adjustments (counter-current heat exchangers); torpedo shaped bodies, reduced appendages, and large size to prevent heat loss.

Marine mammals are able to dive for long periods of time. Both pinnipeds and cetaceans have large and complex blood vessel systems which serve to store oxygen to support deep diving. Other important reservoirs include muscles, blood, and the spleen which all have the capacity to hold a high concentration of oxygen. They are also capable of bradycardia (reduced heart rate), and vasoconstriction (shunting most of the oxygen to vital organs such as the brain and heart) to allow extended diving times and cope with oxygen deprivation. If oxygen is depleted, marine mammals can access substantial reservoirs of glycogen that support anaerobic glycolysis of the cells involved during conditions of systemic hypoxia associated with prolonged submersion.

Sound travels differently through water, and therefore marine mammals have developed adaptations to ensure effective communication, prey capture, and predator detection. The most notable adaptation is the development of echolocation in whales and dolphins. Toothed whales emit a focused beam of high-frequency clicks in the direction that their head is pointing. Sounds are generated by passing air from the bony nares through the phonic lips. These sounds are reflected by the dense concave bone of the cranium and an air sac at its base. The focused beam is modulated by a large fatty organ known as the 'melon'. This acts like an acoustic lens because it is composed of lipids of differing densities.

Marine mammals have evolved a wide variety of features for feeding, which are mainly seen in their dentition. For example, the cheek teeth of pinnipeds and odontocetes are specifically adapted to capture fish and squid. In contrast, baleen whales have evolved baleen plates to filter feed plankton and small fish from the water.

Polar bears, otters, and fur seals have fur, one of the defining mammalian features, that is long, oily, and waterproof in order to trap air to provide insulation. In contrast, other marine mammals – such as whales, dolphins, porpoises, manatees, dugongs, and walruses – have lost long fur in favor of a thick, dense epidermis and a thickened fat layer (blubber) in response to hydrodynamic requirements. Wading and bottom-feeding animals (such as manatees) need to be heavier than water in order to keep contact with the floor or to stay submerged. Surface-living animals (such as sea otters) need the opposite, and free-swimming animals living in open waters (such as dolphins) need to be neutrally buoyant in order to be able to swim up and down the water column. Typically, thick and dense bone is found in bottom feeders and low bone density is associated with mammals living in deep water. Some marine mammals, such as polar bears and otters, have retained four weight-bearing limbs and can walk on land like fully terrestrial animals.

Ecology

Diet

All cetaceans are carnivorous and predatory. Toothed whales mostly feed on fish and cephalopods, followed by crustaceans and bivalves. Some may forage with other kinds of animals, such as other species of whales or certain species of pinnipeds. One common feeding method is herding, where a pod squeezes a school of fish into a small volume, known as a bait ball. Individual members then take turns plowing through the ball, feeding on the stunned fish. Coralling is a method where dolphins chase fish into shallow water to catch them more easily. Killer whales and bottlenose dolphins have also been known to drive their prey onto a beach to feed on it. Other whales with a blunt snout and reduced dentition rely on suction feeding. Though carnivorous, they house gut flora similar to that of terrestrial herbivores, probably a remnant of their herbivorous ancestry.

Killer whale hunting a Weddel seal

Baleen whales use their baleen plates to sieve plankton, among others, out of the water; there are two types of methods: lunge-feeding and gulp-feeding. Lunge-feeders expand the volume of their jaw to a volume bigger than the original volume of the whale itself by inflating their mouth. This causes grooves on their throat to expand, increasing the amount of water the mouth can store. They ram a baitball at high speeds in order to feed, but this is only energy-effective when used against a large baitball. Gulp-feeders swim with an open mouth, filling it with water and prey. Prey must occur in sufficient numbers to trigger the whale's interest, be within a certain size range so that the baleen plates can filter it, and be slow enough so that it cannot escape.

Sea otters have dexterous hands which they use to smash sea urchins off rocks.

Otters are the only marine animals that are capable of lifting and turning over rocks, which they often do with their front paws when searching for prey. The sea otter may pluck snails and other organisms from kelp and dig deep into underwater mud for clams. It is the only marine mammal that catches fish with its forepaws rather than with its teeth. Under each foreleg, sea otters have a loose pouch of skin that extends across the chest which they use to store collected food to bring to the surface. This pouch also holds a rock that is used to break open shellfish and clams, an example of tool use. The sea otters eat while floating on their backs, using their forepaws to tear food apart and bring to their mouths. Marine otters mainly feed on crustaceans and fish.

Pinnipeds mostly feed on fish and cephalopods, followed by crustaceans and bivalves, and then zooplankton and warm-blooded prey (like sea birds). Most species are generalist feeders, but a few are specialists. They typically hunt non-schooling fish, slow-moving or immobile invertebrates or endothermic prey when in groups. Solitary foraging species usually exploit coastal waters, bays and rivers. When large schools of fish or squid are available, pinnipeds hunt cooperatively in large groups, locating and herding their prey. Some species, such as California and South American sea lions, may forage with cetaceans and sea birds.

The polar bear is the most carnivorous species of bear, and its diet primarily consists of ringed (*Pusa hispida*) and bearded (*Erignathus barbatus*) seals. Polar bears hunt primarily at the interface between ice, water, and air; they only rarely catch seals on land or in open water. The polar bear's most common hunting method is still-hunting: The bear locates a seal breathing hole using its sense of smell, and crouches nearby for a seal to appear. When the seal exhales, the bear smells its breath, reaches into the hole with a forepaw, and drags it out onto the ice. The polar bear also hunts by stalking seals resting on the ice. Upon spotting a seal, it walks to within 100 yards (90 m), and then crouches. If the seal does not notice, the bear creeps to within 30 to 40 feet (9 to 10 m) of the seal and then suddenly rushes to attack. A third hunting method is to raid the birth lairs that female seals create in the snow. They may also feed on fish.

A dugong feeding on the sea-floor

Sirenians are referred to as "sea cows" because their diet consists mainly of sea-grass. When eating, they ingest the whole plant, including the roots, although when this is impossible they feed on just the leaves. A wide variety of seagrass has been found in dugong stomach contents, and evidence exists they will eat algae when seagrass is scarce. West Indian manatees eat up to 60 different species of plants, as well as fish and small invertebrates to a lesser extent.

Keystone Species

Sea otters are a classic example of a keystone species; their presence affects the ecosystem more profoundly than their size and numbers would suggest. They keep the population of certain benthic (sea floor) herbivores, particularly sea urchins, in check. Sea urchins graze on the lower stems of kelp, causing the kelp to drift away and die. Loss of the habitat and nutrients provided by kelp forests leads to profound cascade effects on the marine ecosystem. North Pacific areas that do not have sea otters often turn into urchin barrens, with abundant sea urchins and no kelp forest. Reintroduction of sea otters to British Columbia has led to a dramatic improvement in the health of coastal ecosystems, and similar changes have been observed as sea otter populations recovered in the Aleutian and Commander Islands and the Big Sur coast of California However, some kelp forest ecosystems in California have also thrived without sea otters, with sea urchin populations apparently controlled by other factors. The role of sea otters in maintaining kelp forests has been observed to be more important in areas of open coast than in more protected bays and estuaries.

Antarctic fur seal pups vs. Arctic harp seal pup

An apex predator affects prey population dynamics and defense tactics (such as camouflage). The polar bear is the apex predator within its range. Several animal species, particularly Arctic foxes (*Vulpes lagopus*) and glaucous gulls (*Larus hyperboreus*), routinely scavenge polar bear kills. The relationship between ringed seals and polar bears is so close that the abundance of ringed seals in some areas appears to regulate the density of polar bears, while polar bear predation in turn regulates density and reproductive success of ringed seals. The evolutionary pressure of polar bear predation on seals probably accounts for some significant differences between Arctic and Antarctic seals. Compared to the Antarctic, where there is no major surface predator, Arctic seals use more breathing holes per individual, appear more restless when hauled out on the ice, and rarely defecate on the ice. The fur of Arctic pups is white, presumably to provide camouflage from predators, whereas Antarctic pups all have dark fur.

Killer whales are apex predators throughout their global distribution, and can have a profound effect on the behavior and population of prey species. Their diet is very broad and they can feed on many vertebrates in the ocean including salmon, rays, sharks (even white sharks), large baleen whales, and nearly 20 species of pinniped. The predation of whale calves may be responsible for annual whale migrations to calving grounds in more tropical waters, where the population of killer whales is much lower than in polar waters. Prior to whaling, it is thought that great whales were a major food source; however, after their sharp decline, killer whales have since expanded their diet, leading to the decline of smaller marine mammals. A decline in Aleutian Islands sea otter populations in the 1990s was controversially attributed by some scientists to killer whale predation, although with no direct evidence. The decline of sea otters followed a decline in harbor seal and Steller sea lion populations, the killer whale's preferred prey, which in turn may be substitutes for their original prey, now reduced by industrial whaling.

Whale Pump

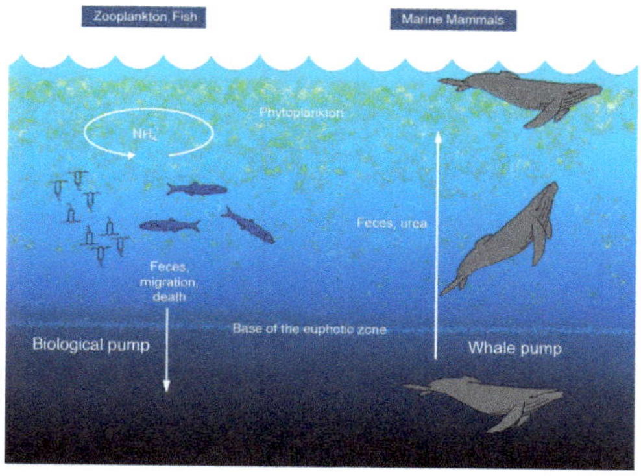

"Whale pump" – the role played by whales in recycling ocean nutrients

A 2010 study considered whales to be a positive influence to the productivity of ocean fisheries, in what has been termed a "whale pump". Whales carry nutrients such as nitrogen from the depths back to the surface. This functions as an upward biological pump, reversing an earlier presumption that whales accelerate the loss of nutrients to the bottom. This nitrogen input in the Gulf of Maine is more than the input of all rivers combined emptying into the gulf, some 25,000 short tons (23,000 t) each year. Whales defecate at the ocean's surface; their excrement is important for fisheries because it is rich in iron and nitrogen. The whale feces are liquid and instead of sinking, they stay at the surface where phytoplankton feed off it.

Upon death, whale carcasses fall to the deep ocean and provide a substantial habitat for marine life. Evidence of whale falls in present-day and fossil records shows that deep sea whale falls support a rich assemblage of creatures, with a global diversity of 407 species, comparable to other neritic biodiversity hotspots, such as cold seeps and hydrothermal vents. Deterioration of whale carcasses happens though a series of three stages. Initially, moving organisms, such as sharks and hagfish, scavenge soft tissue at a rapid rate over a period of months to as long as two years. This is followed by the colonization of bones and surrounding sediments (which contain organic matter) by enrichment opportunists, such as crustaceans and polychaetes, throughout a period of years. Finally, sulfophilic bacteria reduce the bones releasing hydrogen sulphide enabling the growth of chemoautotrophic organisms, which in turn, support other organisms such as mussels, clams, limpets, and sea snails. This stage may last for decades and supports a rich assemblage of species, averaging 185 species per site.

Interactions with Humans

Threats

Exploitation

Men killing northern fur seals on Saint Paul Island, Alaska in the 1890s

Marine mammals were hunted by coastal aboriginal humans historically for food and other resources. These subsistence hunts still occur in Canada, Greenland, Indonesia, Russia, the United States, and several nations in the Caribbean. The effects of these are only localized, as hunting efforts were on a relatively small scale. Commercial hunting took this to a much greater scale and marine mammals were heavily exploited. This led to the extinction of the Steller's sea cow (along with subsistence hunting) and the Caribbean monk seal. Today, populations of species that were historically hunted, such as blue whales (*Balaenoptera musculus*) and the North Pacific right whale (*Eubalaena japonica*), are much lower than their pre-whaling levels. Because whales generally have slow growth rates, are slow to reach sexual maturity, and have a low reproductive output, population recovery has been very slow.

A number of whales are still subject to direct hunting, despite the 1986 moratorium ban on whaling set under the terms of the International Whaling Commission (IWC). There are only two nations remaining which sanction commercial whaling: Norway, where several hundred common minke whales are harvested each year; and Iceland, where quotas of 150 fin whales and 100 minke whales per year are set. Japan also harvests several hundred Antarctic and North Pacific minke whales each year, ostensibly for scientific research in accordance with the moratorium. However, the illegal trade of whale and dolphin meat is a significant market in Japan and some countries.

The most profitable furs in the fur trade were those of sea otters, especially the northern sea otter which inhabited the coastal waters between the Columbia River to the south and Cook Inlet to the north. The fur of the Californian southern sea otter was less highly prized and thus less profitable. After the northern sea otter was hunted to local extinction, maritime fur traders shifted to California until the southern sea otter was likewise nearly extinct. The British and American maritime fur traders took their furs to the Chinese port of Guangzhou (Canton), where they worked within the established Canton System. Furs from Russian America were mostly sold to China via the Mongolian trading town of Kyakhta, which had been opened to Russian trade by the 1727 Treaty of Kyakhta.

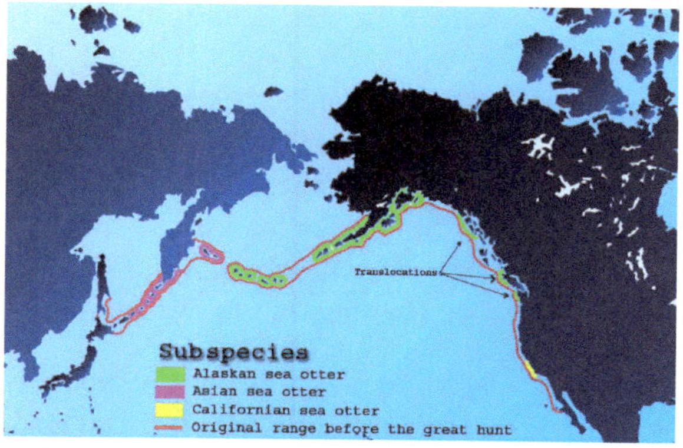

Historical and modern range of northern sea otters

Commercial sealing was historically just as important as the whaling industry. Exploited species included harp seals, hooded seals, Caspian seals, elephant seals, walruses and all species of fur seal. The scale of seal harvesting decreased substantially after the 1960s, after the Canadian government reduced the length of the hunting season and implemented measures to protect adult females. Several species that were commercially exploited have rebounded in numbers; for example, Antarctic fur seals may be as numerous as they were prior to harvesting. The northern elephant seal was hunted to near extinction in the late 19th century, with only a small population remaining on Guadalupe Island. It has since recolonized much of its historic range, but has a population bottleneck. Conversely, the Mediterranean monk seal was extirpated from much of its former range, which stretched from the Mediterranean to the Black Sea and northwest Africa, and only remains in the northeastern Mediterranean and some parts of northwest Africa.

Polar bears can be hunted for sport in Canada with a special permit and accompaniment by a local guide. This can be an important source of income for small communities, as guided hunts bring in more income than selling the polar bear hide on markets. The United States, Russia, Norway, Greenland, and Canada allow subsistence hunting, and Canada distributes hunting permits to indigenous communities. The selling of these permits is a main source of income for many of these communities. Their hides can be used for subsistence purposes, kept as hunting trophies, or can be bought in markets.

Ocean Traffic and Fisheries

The remains of a North Atlantic right whale after it collided with a ship propeller

By-catch is the incidental capture of non-target species in fisheries. Fixed and drift gill nets cause the highest mortality levels for both cetaceans and pinnipeds, however, entanglements in long lines, mid-water trawls, and both trap and pot lines are also common. Tuna seines are particularly problematic for entanglement by dolphins. By-catch affects all cetaceans, both small and big, in all habitat types. However, smaller cetaceans and pinnipeds are most vulnerable as their size means that escape once they are entangled is highly unlikely and they frequently drown. While larger cetaceans are capable of dragging nets with them, the nets sometimes remain tightly attached to the individual and can impede the animal from feeding sometimes leading to starvation. Abandoned or lost nets and lines cause mortality through ingestion or entanglement. Marine mammals also get entangled in aquaculture nets, however, these are rare events and not prevalent enough to impact populations.

Vessel strikes cause death for a number of marine mammals, especially whales. In particular, fast commercial vessels such as container ships can cause major injuries or death when they collide with marine mammals. Collisions occur both with large commercial vessels and recreational boats and cause injury to whales or smaller cetaceans. The critically endangered North Atlantic right whale is particularly affected by vessel strikes. Tourism boats designed for whale and dolphin watching can also negatively impact on marine mammals by interfering with their natural behavior.

The fishery industry not only threatens marine mammals through by-catch, but also through competition for food. Large scale fisheries have led to the depletion of fish stocks that are important prey species for marine mammals. Pinnipeds have been especially affected by the direct loss of food supplies and in some cases the harvesting of fish has led to food shortages or dietary deficiencies, starvation of young, and reduced recruitment into the population. As the fish stocks have been depleted, the competition between marine mammals and fisheries has sometimes led to conflict. Large-scale culling of populations of marine mammals by commercial fishers has been initiated in a number of areas in order to protect fish stocks for human consumption.

Shellfish aquaculture takes up space so in effect creates competition for space. However, there is little direct competition for aquaculture shellfish harvest. On the other hand, marine mammals regularly take finfish from farms, which creates significant problems for marine farmers. While there are usually legal mechanisms designed to deter marine mammals, such as anti-predator nets or harassment devices, individuals are often illegally shot.

Habitat Loss and Degradation

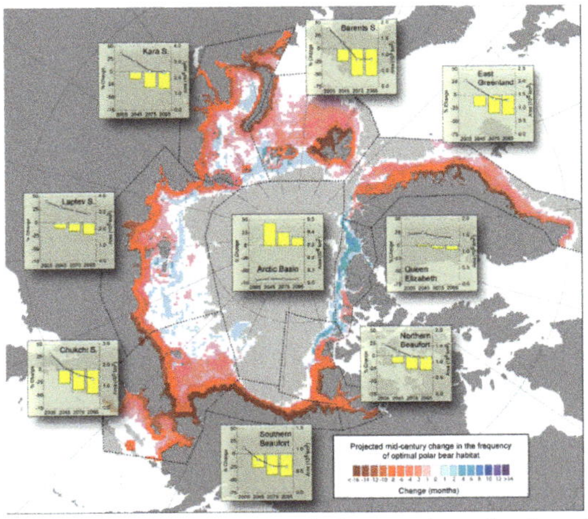

Map from the U.S. Geological Survey shows projected changes in polar bear habitat from 2005 to 2095. Red areas indicate loss of optimal polar bear habitat; blue areas indicate gain.

Contaminants that are discharged into the marine environment accumulate in the bodies of marine mammals when they are stored unintentionally in their blubber along with energy. Contaminants that are found in the tissues of marine mammals include heavy metals, such as mercury and lead, but also organochlorides and polycyclic aromatic hydrocarbons. For example, these can cause disruptive effects on endocrine systems; impair the reproductive system, and lower the immune system of individuals, leading to a higher number of deaths. Other pollutants such as oil, plastic debris and sewage threaten the livelihood of marine mammals.

Noise pollution from anthropogenic activities is another major concern for marine mammals. This is a problem because underwater noise pollution interferes with the abilities of some marine mammals to communicate, and locate both predators and prey. Underwater explosions are used for a variety of purposes including military activities, construction and oceanographic or geophysical research. They can cause injuries such as hemorrhaging of the lungs, and contusion and ulceration of the gastrointestinal tract. Underwater noise is generated from shipping, the oil and gas industry, research, and military use of sonar and oceanographic acoustic experimentation. Acoustic harassment devices and acoustic deterrent devices used by aquaculture facilities to scare away marine mammals emit loud and noxious underwater sounds.

Two changes to the global atmosphere due to anthropogenic activity threaten marine mammals. The first is increases in ultraviolet radiation due to ozone depletion, and this mainly affects the Antarctic and other areas of the southern hemisphere. An increase in ultraviolet radiation has the capacity to decrease phytoplankton abundance, which forms the basis of the food chain in the ocean. The second effect of global climate change is global warming due to increased carbon dioxide levels in the atmosphere. Raised sea levels, sea temperature and changed currents are expected to affect marine mammals by altering the distribution of important prey species, and changing the suitability of breeding sites and migratory routes. The Arctic food chain would be disrupted by the near extinction or migration of polar bears. Arctic sea ice is the polar bear's habitat. It has been declining at a rate of 13% per decade because the temperature is rising at twice the rate of the rest of the world. By the year 2050, up to two-thirds of the world's polar bears may vanish if the sea ice continues to melt at its current rate.

Protection

Logo for the International Whaling Commission

The Marine Mammal Protection Act of 1972 (MMPA) was passed on October 21, 1972 under president Richard Nixon to prevent the further depletion and possible extinction of marine mammal stocks. It prohibits the taking ("the act of hunting, killing, capture, and/or harassment of any marine mammal; or, the attempt at such") of any marine mammal without a permit issued by the Secretary. Authority to manage the MMPA was divided between the Secretary of the Interior through the U.S. Fish and Wildlife Service (Service), and the Secretary of Commerce, which is delegated to the National Oceanic and Atmospheric Administration (NOAA). The Marine Mammal Commission (MMC) was established to review existing policies and make recommendations to the Service and NOAA to better implement the MMPA. The Service is responsible for ensuring the protection of sea otters and marine otters, walruses, polar bears, the three species of manatees, and dugongs; and NOAA was given responsibility to conserve and manage pinnipeds (excluding walruses) and cetaceans.

The 1979 Convention on the Conservation of Migratory Species of Wild Animals (CMS) is the only global organization that conserves a broad range of animals, of which includes marine mammals. Of the agreements made, three of them deal with the conservation of marine mammals: ACCOBAMS, ASCOBANS, and the Wadden Sea Agreement. In 1982, the United Nations Convention on the Law of the Sea (LOSC) adopted a pollution prevention approach to conservation, which many other conventions at the time also adopted.

The Agreement on the Conservation of Cetaceans in the Black Sea, Mediterranean Sea and contiguous Atlantic area (ACCOBAMS), founded in 1996, specifically protects cetaceans in the Mediterranean area, and "maintains a favorable status", a direct action against whaling. There are 23 member states. The Agreement on the Conservation of Small Cetaceans of the Baltic and North Seas (ASCOBANS) was adopted alongside ACCOBAMS to establish a special protection area for Europe's increasingly threatened cetaceans. Other anti-whaling efforts include a ten-year moratorium in 1986 by the IWC on all whaling, and an environmental agreement (a type of international law) the International Convention for the Regulation of Whaling which controlled commercial, scientific, and subsistence whaling.

The Agreement on the Conservation of Seals in the Wadden Sea, enforced in 1991, prohibits the killing or harassment of seals in the Wadden Sea, specifically targeting the harbor seal population.

The 1973 Agreement on the Conservation of Polar Bears between Canada, Denmark (Greenland), Norway (Svalbard), the United States, and the Soviet Union outlawed the unregulated hunting of polar bears from aircraft and icebreakers, as well as protecting migration, feeding, and hibernation sites.

Various non-governmental organizations participate in marine conservation activism, wherein they draw attention to and aid in various problems in marine conservation, such as pollution, whaling, bycatch, and so forth. Notable organizations include the Greenpeace who focus on overfishing and whaling among other things, and Sea Shepherd Conservation Society who are known for taking direct-action tactics to expose illegal activity.

As Food

For thousands of years, indigenous peoples of the Arctic have depended on whale meat. The meat is harvested from legal, non-commercial hunts that occur twice a year in the spring and autumn. The meat is stored and eaten throughout the winter. The skin and blubber (muktuk) taken from the bowhead, beluga, or narwhal is also valued, and is eaten raw or cooked. Whaling has also been practiced in the Faroe Islands in the North Atlantic since about the time of the first Norse settlements on the islands. Around 1000 long-finned pilot whales are still killed annually, mainly during the summer. Today, dolphin meat is consumed in a small number of countries worldwide, which include Japan and Peru (where it is referred to as *chancho marino*, or "sea pork"). In some parts of the world, such as Taiji, Japan and the Faroe Islands, dolphins are traditionally considered food, and are killed in harpoon or drive hunts.

Pilot whale meat (bottom), blubber (middle) and dried fish (left) with potatoes, Faroe Islands

There have been human health concerns associated with the consumption of dolphin meat in Japan after tests showed that dolphin meat contained high levels of methylmercury. There are no known cases of mercury poisoning as a result of consuming dolphin meat, though the government continues to monitor people in areas where dolphin meat consumption is high. The Japanese government recommends that children and pregnant women avoid eating dolphin meat on a regular basis. Similar concerns exist with the consumption of dolphin meat in the Faroe Islands, where prenatal exposure to methylmercury and PCBs primarily from the consumption of pilot whale meat has resulted in neuropsychological deficits amongst children.

"The Faroe Islands population was exposed to methylmercury largely from contaminated pilot whale meat, which contained very high levels of about 2 mg methylmercury/kg. However, the Faroe Islands populations also eat significant amounts of fish. The study of about 900 Faroese children showed that prenatal exposure to methylmercury resulted in neuropsychological deficits at 7 years of age."

— *World Health Organization*

Ringed seals were once the main food staple for the Inuit. They are still an important food source for the people of Nunavut and are also hunted and eaten in Alaska. Seal meat is an important source of food for residents of small coastal communities. The seal blubber is used to make seal oil, which is marketed as a fish oil supplement. In 2001, two percent of Canada's raw seal oil was processed and sold in Canadian health stores.

In Captivity

Aquariums

Cetaceans

Various species of dolphins are kept in captivity. These small cetaceans are more often than not kept in theme parks and dolphinariums, such as SeaWorld. Bottlenose dolphins are the most common species of dolphin kept in dolphinariums as they are relatively easy to train and have a long lifespan in captivity. Hundreds of bottlenose dolphins live in captivity across the world, though exact numbers are hard to determine. The dolphin "smile" makes them popular attractions, as this is a welcoming facial expression in humans; however the smile is due to a lack of facial muscles and subsequent lack of facial expressions.

Performing killer whale at SeaWorld San Diego, 2009

Organizations such as World Animal Protection and the Whale and Dolphin Conservation Society campaign against the practice of keeping cetaceans, particularly killer whales, in captivity. In captivity, they often develop pathologies, such as the dorsal fin collapse seen in 60–90% of male killer whales. Captives have vastly reduced life expectancies, on average only living into their 20s. In the wild, females who survive infancy live 46 years on average, and up to 70–80 years in rare cases. Wild males who survive infancy live 31 years on average, and up to 50–60 years. Captivity usually bears little resemblance to wild habitat, and captive whales' social groups are foreign to those found in the wild. Captive life is also stressful due the requirement to perform circus tricks that are not part of wild killer whale behavior, as well as restricting pool size. Wild killer whales may travel up to 100 miles (160 km) in a day, and critics say the animals are too big and intelligent to be suitable for captivity. Captives occasionally act aggressively towards themselves, their tankmates, or humans, which critics say is a result of stress. Dolphins are often trained to do several anthropomorphic behaviors, including waving and kissing—behaviors wild dolphins would rarely do.

Pinnipeds

A sea lion trained to balance a ball on its nose

The large size and playfulness of pinnipeds make them popular attractions. Some exhibits have rocky backgrounds with artificial haul-out sites and a pool, while others have pens with small rocky, elevated shelters where the animals can dive into their pools. More elaborate exhibits contain deep pools that can be viewed underwater with rock-mimicking cement as haul-out areas. The most common pinniped species kept in captivity is the California sea lion as it is abundant and easy to train. These animals are used to perform tricks and entertain visitors. Other species popularly kept in captivity include the grey seal and harbor seal. Larger animals like walruses and Steller sea lions are much less common. Pinnipeds are popular attractions because they are "disneyfied", and, consequently, people often anthropomorphize them with a curious, funny, or playful nature.

Some organizations, such as the Humane Society of the United States and World Animal Protection, object to keeping pinnipeds and other marine mammals in captivity. They state that the exhibits could not be large enough to house animals that have evolved to be migratory, and a pool could never replace the size and biodiversity of the ocean. They also oppose using sea lions for entertainment, claiming the tricks performed are "exaggerated variations of their natural behaviors" and distract the audience from the animal's unnatural environment.

Otter

Sea otters can do well in captivity, and are featured in over 40 public aquariums and zoos. The Seattle Aquarium became the first institution to raise sea otters from conception to adulthood with the birth of Tichuk in 1979, followed by three more pups in the early 1980s. In 2007, a YouTube video of two cute sea otters holding paws drew 1.5 million viewers in two weeks, and had over 20 million views as of January 2015. Filmed five years previously at the Vancouver Aquarium, it was YouTube's most popular animal video at the time, although it has since been surpassed. Otters are often viewed as having a "happy family life", but this is an anthropomorphism.

Sirenians

The oldest manatee in captivity is Snooty, at the South Florida Museum's Parker Manatee Aquarium in Bradenton, Florida. Born at the Miami Aquarium and Tackle Company on July 21, 1948, Snooty was one of the first recorded captive manatee births. He was raised entirely in captivity, and will never be released into the wild. Manatees can also be viewed in a number of European zoos, such as the Tierpark in Berlin, the Nuremberg Zoo, in ZooParc de Beauval in France, and in the Aquarium of Genoa in Italy. The River Safari at Singapore features seven of them.

Military

Bottlenose dolphins and California sea lions were used in the United States Navy Marine Mammal Program (NMMP) to detect mines, protect ships from enemy soldiers, and recover objects. The Navy has never trained attack dolphins, as they would not be able to discern allied soldiers from enemy soldiers. There were five marine mammal teams, each purposed for one of the three tasks: MK4 (dolphins), MK5 (sea lions), MK6 (dolphins and sea lions), MK7 (dolphins), and MK8 (dolphins); MK is short for mark. The dolphin teams were trained to detect and mark mines either attached to the seafloor or floating in the water column, because dolphins can use their echolocative abilities to detect mines. The sea lion team retrieved test equipment such as fake mines or bombs dropped from planes usually out of reach of divers who would have to make multiple dives. MK6 protects harbors and ships from enemy divers, and was operational in the Gulf War and Vietnam

War. The dolphins would swim up behind enemy divers and attach a buoy to their air tank, so that they would float to the surface and alert nearby Navy personnel. Sea lions would hand-cuff the enemy, and try to outmaneuver their counter-attacks.

A dolphin wearing a locating pinger, performing mine clearance work in the Iraq War

The use of marine mammals by the Navy, even in accordance with the Navy's policy, continues to meet opposition. The Navy's policy says that only positive reinforcement is to be used while training the military dolphins, and that they be cared for in accordance with accepted standards in animal care. The inevitable stresses involved in training are topics of controversy, as their treatment is unlike the animals' natural lifestyle, especially towards their confined spaces when not training. There is also controversy over the use of muzzles and other inhibitors, which prevent the dolphins from foraging for food while working. The Navy states that this is to prevent them from ingesting harmful objects, but conservation activists say this is done to reinforce the trainers' control over the dolphins, who hand out food rewards. The means of transportation is also an issue for conservation activists, since they are hauled in dry carriers, and switching tanks and introducing the dolphin to new dolphins is potentially dangerous as they are territorial.

Baikal Seal

The Baikal seal, Lake Baikal seal or *nerpa* (*Pusa sibirica*), is a species of earless seal endemic to Lake Baikal in Siberia, Russia. Like the Caspian seal, it is related to the Arctic ringed seal. The Baikal seal is one of the smallest true seals and the only exclusively freshwater pinniped species. A subpopulation of inland harbour seals living in the Hudson's Bay region of Quebec, Canada (*lac de loups marins* harbour seals), the Saimaa ringed seal (a ringed seal subspecies) and the Ladoga seal (a ringed seal subspecies) are found in fresh water, but these are part of species that also have marine populations. It remains a scientific mystery how the seals originally came to Lake Baikal, hundreds of kilometers from any ocean.

The most recent population estimates are 80,000-100,000 animals, roughly equaling the expected carrying capacity of the lake. At present, the species is not considered threatened, despite hunting (both legal and illegal) and heavy pollution of the lake.

Description

The Baikal seal is one of the smallest true seals. Adults grow to around 1.3 m (4 ft 3 in) in length with a body mass from 63 to 70 kg (139 to 154 lb). The animals show very little sexual dimorphism; males are only slightly larger than females. They have a uniform, steely-grey coat on their backs and fur with a yellowish tinge on their abdomens. As the coat weathers, it becomes brownish. When first born, pups weigh around 4.5 kg (9.9 lb) and have coats of white, silky, natal fur. This fur is quickly shed and exchanged for a darker coat, much like that of adults. Rarely, Baikal seals can be found with spotted coats.

A young seal

Distribution

The Baikal seal lives only in the waters of Lake Baikal. It is something of a mystery how Baikal seals came to live there in the first place. They may have swum up rivers and streams or possibly Lake Baikal was linked to the ocean at some point through a large body of water, such as the West Siberian Glacial Lake or West Siberian Plain, formed in a previous ice age. The seals are estimated to have inhabited Lake Baikal for some two million years.

The areas of the lake in which the Baikal seals reside change depending on the season, as well as other environmental factors. They are solitary animals for the majority of the year, sometimes living kilometers away from other Baikal seals. In general, a higher concentration of Baikal seals is found in the northern parts of the lake, because the longer winter keeps the ice frozen longer, which is preferable for pupping. However, in recent years, migrations to the southern half of the lake have occurred, possibly to evade hunters. In winter, when the lake is frozen over, seals maintain a few breathing holes over a given area and tend to remain nearby, not interfering with the food supplies of nearby seals. When the ice begins to melt, Baikal seals tend to keep to the shoreline.

Abundance and Trends

Since 2008, the Baikal seal has been listed as a Least Concern species on the IUCN Red List. This means that they are not currently threatened or endangered. In 1994, the Russian government estimated that they numbered 104,000. In 2000, Greenpeace performed its own count and found an estimated 55,000 to 65,000 seals. The most recent estimates are 80,000-100,000 animals, roughly equaling the carrying capacity of the lake.

In the last century, the kill quota for hunting Baikal seals was raised several times, most notably after the fur industry boomed in the late 1970s and when official counts began indicating more Baikal seals were present than previously known. The quota in 1999, 6,000, was lowered in 2000 to 3,500, which is still nearly 5% of the population if the Greenpeace count is correct. In addition, new techniques, such as netting breathing holes and seal dens to catch pups, have been introduced. One prime seal pelt will bring 1,000 rubles at market. In 2004-2006, about 2,000 seals were killed per year according to official Russian statistics, but another 1,500-4,000 are thought to die annually due to drowning in fishing gear, poaching, and the like.

A Baikal seal costume at the Nerpa Festival in Irkutsk, Baikal region, Russia

Lake Baikal has eight wildlife patrol officers, which amounts to one officer for roughly 2,500 square kilometers, making enforcement of regulations difficult. Even without poaching, hunting, even on a small scale, is a problem, because many of the seals that are shot or injured still escape and die later. These do not fall under the kill quota and are added after. Poaching and hunting are unlikely to slow considerably without government intervention.

Another problem at Lake Baikal is the introduction of pollutants into the ecosystem. Pesticides such as DDT and hexachlorocyclohexane, as well as industrial waste, mainly from the Baikalisk pulp and paper plant, are thought to have been the cause of several disease epidemics among Baikal seal populations. The chemicals are speculated to concentrate up the food chain and weaken the Baikal seal's immune system, making them susceptible to diseases such as canine distemper and the plague, which was the cause of a serious Baikal seal epidemic that resulted in the deaths of thousands of animals in 1997 and 1999. Baikal seal pups have higher levels of DDT and PCB than known in any other population of European or Arctic earless seal.

The most serious future threat to the survival of the seal may be global warming, which has the potential to seriously affect a closed cold-water ecosystem such as that of Lake Baikal.

The only known natural predator of this seal is the brown bear, but this is not believed to occur frequently.

Reproduction

Female Baikal seals reach sexual maturity at 3–6 years of age, whereas males achieve it around 4–7 years. The males and females are not strongly sexually dimorphic; this fact, compounded by their poor vision and the further optical distortions produced by cold water, occasionally leads males to mistakenly attempt to copulate with other males, a behavior formerly erroneously interpreted by ethologists as homosexual. Baikal seals mate in the water towards the end of the pupping season. With a combination of delayed implantation and a nine-month gestation period, the Baikal seals' overall pregnancy is around 11 months. Pregnant females are the only Baikal seals to haul out during the winter. The males tend to stay in the water, under the ice, all winter. Females usually give birth to one pup, but they are one of only two species of true seals with the ability to give birth to twins. The twins often stick together for some time after being weaned. The females, after giving birth to their pups on the ice in late winter, become immediately impregnated again, and often are lactating while pregnant.

Baikal seals are slightly polygamous and slightly territorial, although not particularly defensive of their territory. Males mate with around three females if given the chance. They then mark the female's den with a strong, musky odor, which can be smelled by another male if he approaches. The female raises the pups on her own; she digs them a fairly large den under the ice, up to 5 m (16 ft) in length, and more than 2 m (6 ft) wide. Pups as young as two days old then further expand this den by digging a maze of tunnels around the den. Since the pup avoids breaking the surface with these tunnels, this activity is thought to be mainly for exercise, to keep warm until they have built up an insulating layer of blubber.

The mother Baikal seal feeds her young for around 2.5 months, nearly twice as long as any other seal. During this time, the pups can increase their birth weight (around 4 kg {9 lb}) five-fold. After the pups are weaned, the mother introduces them to solid food, bringing shrimp, fish, and other food into the den.

In spring, when the ice melts and the dens usually collapse, the pup is left to fend for itself. Growth continues until they are 20 to 25 years old.

Every year in the late winter and spring, both sexes haul themselves out and begin to moult their coat from the previous year, which is replaced with new fur. While moulting, they refrain from eating and enter a lethargic state, during which time they often die of overheating, males especially, from lying on the ice too long in the sun. During the spring and summer, groups as large as 500 can form on the ice floes and shores of Lake Baikal. Baikal seals can live to over 50 years old, exceptionally old for a seal, although they are presumed to be fertile only until they are around 40.

Foraging

Their main food source is the golomyanka, a cottoid oilfish found only in Lake Baikal. Baikal seals eat more than half of the annual produced biomass of golomyanka, some 64,000 tons. They also eat some types of invertebrates, and the occasional omul. They feed mainly at night, when fish come within 100 m (330 ft) of the surface. They feed during 10-20 minute dives, although this is hardly the extent of their submarine abilities. Baikal seals have two litres more blood than any other seal of their size and can stay underwater for up to 70 minutes if they are frightened or need to escape danger.

The Baikal seal has been blamed for drops in omul numbers, but this is not the case. The omul's main competitor is the golomyanka and by eating tons of these fish a year, Baikal seals cut down on the omul's competition for resources.

Baikal seals have one unusual foraging habit. In early autumn, before the entire lake freezes over, they migrate to bays and coves and hunt Kessler's sculpin, a fish that lives in silty areas and, as a result, usually contains grit and silt in its digestive system. This grit scours the seals' gastrointestinal tracts and expels parasites.

Whale

Whales are a widely distributed and diverse group of fully aquatic placental marine mammals. They are an informal grouping within the infraorder Cetacea, usually excluding dolphins and porpoises. Whales, dolphins and porpoises belong to the order Cetartiodactyla with even-toed ungulates and their closest living relatives are the hippopotamuses, having diverged about 40 million years ago. The two parvorders of whales, baleen whales (Mysticeti) and toothed whales (Odontoceti), are thought to have split apart around 34 million years ago. The whales comprise eight extant families: Balaenopteridae (the rorquals), Balaenidae (right whales), Cetotheriidae (the pygmy right whale), Eschrichtiidae (the grey whale), Monodontidae (belugas and narwhals), Physeteridae (the sperm whale), Kogiidae (the dwarf and pygmy sperm whale), and Ziphiidae (the beaked whales).

Whales are creatures of the open ocean; they feed, mate, give birth, suckle and raise their young at sea. So extreme is their adaptation to life underwater that they are unable to survive on land. Whales range in size from the 2.6 metres (8.5 ft) and 135 kilograms (298 lb) dwarf sperm whale to the 29.9 metres (98 ft) and 190 metric tons (210 short tons) blue whale, which is the largest creature that has ever lived. The sperm whale is the largest toothed predator on earth. Several species exhibit sexual dimorphism, in that the females are larger than males. Baleen whales have no teeth; instead they have plates of baleen, a fringe-like structure used to expel water while retaining the krill and plankton which they feed on. They use their throat pleats to expand the mouth to take in huge gulps of water. Balaenids have heads that can make up 40% of their body mass to take in water. Toothed whales, on the other hand, have conical teeth designed for catching fish or squid. Baleen whales have a well developed sense of "smell", whereas toothed whales have well-developed hearing – their hearing, that is adapted for both air and water, is so well developed that some can survive even if they are blind. Some species, such as sperm whales, are well adapted for diving to great depths to catch squid and other favoured prey.

Whales have evolved from land-living mammals. As such they must breathe air regularly, though they can remain submerged for long periods. They have blowholes (modified nostrils) located on top of their heads, through which air is taken in and expelled. They are warm-blooded, and have a layer of fat, or blubber, under the skin. With streamlined fusiform bodies and two limbs that are modified into flippers, whales can travel at up to 20 knots, though they are not as flexible or agile as seals. Whales produce a great variety of vocalizations, notably the extended songs of the humpback whale. Although whales are widespread, most species prefer the colder waters of the Northern and Southern Hemispheres, and migrate to the equator to give birth. Species such as humpbacks and blue whales are capable of travelling thousands of miles without feeding. Males typically mate with multiple females every year, but females only mate every two to three years. Calves are typically born in the spring and summer months and females bear all the responsibility for raising them. Mothers of some species fast and nurse their young for one to two years.

Once relentlessly hunted for their products, whales are now protected by international law. The North Atlantic right whales nearly became extinct in the twentieth century, with a population low of 450, and the North Pacific grey whale population is ranked Critically Endangered by the IUCN. Besides whaling, they also face threats from bycatch and marine pollution. The meat, blubber and baleen of whales have traditionally been used by indigenous peoples of the Arctic. Whales have been depicted in various cultures worldwide, notably by the Inuit and the coastal peoples of Vietnam and Ghana, who sometimes hold whale funerals. Whales occasionally feature in literature and film, as in the great white whale of Herman Melville's *Moby Dick*. Small whales, such as belugas, are sometimes kept in captivity and trained to perform tricks, but breeding success has been poor and the animals often die within a few months of capture. Whale watching has become a form of tourism around the world.

Etymology and Definitions

The word "whale" comes from the Old English *whæl*, related to the High German *wal*. Related forms are the Old Norse *hvalr* and the Swedish/Danish *hval*. The obsolete "whalefish" has a similar derivation, indicating a time when whales were thought to be fish. Other archaic English forms include *wal, wale, whal, whalle, whaille, wheal*, etc.

The term "whale" is sometimes used interchangeably with dolphins and porpoises, acting as a synonym for Cetacea. Six species of dolphins have the word "whale" in their name, collectively known as blackfish: the killer whale, the melon-headed whale, the pygmy killer whale, the false killer whale, and the two species of pilot whales, all of which are classified under the family Delphinidae (oceanic dolphins). Each species has a different reason for it, for example, the killer whale was named "Ballena asesina" by Spanish sailors, which translates directly to "whale assassin" or "whale killer", but is more often translated to "killer whale".

Taxonomy and Evolution

The hippopotamuses are the sister clade to the Cetacea, which contains whales, dolphins and porpoises.

The whales are part of the largely terrestrial mammalian clade Laurasiatheria. Whales do not form a clade or order; the infraorder Cetacea includes dolphins and porpoises, which are not considered whales.

Cetaceans are divided into two parvorders: the largest parvorder, Mysticeti (baleen whales), is characterized by the presence of baleen, a sieve-like structure in the upper jaw made of keratin, which it uses to filter plankton, among others, from the water; Odontocetes (toothed whales) are characterized by bearing sharp teeth for hunting, as opposed to their counterparts' baleen.

Cetaceans and artiodactyls now are classified under the order Cetartiodactyla, often still referred to as Artiodactyla, which includes both whales and hippopotamuses. The hippopotamus and pygmy hippopotamus are the whale's closest terrestrial living relatives.

Mysticetes

Mysticetes are also known as baleen whales. They have a pair of blowholes side-by-side and lack teeth, which renders them incapable of catching larger prey; they instead have baleen plates which is a sieve-like structure in the upper jaw made of keratin, which they use to filter plankton and other food from the water; this forces them to follow krill or plankton migrations. Some whales, such as the humpback, reside in the polar regions where they feed on a reliable source of schooling fish and krill. These animals rely on their well-developed flippers and tail fin to propel themselves through the water; they swim by moving their fore-flippers and tail fin up and down. Whale ribs loosely articulate with their thoracic vertebrae at the proximal end, but do not form a rigid rib cage. This adaptation allows the chest to compress during deep dives as the pressure increases. Mysticetes consist of four families: rorquals (balaenopterids), cetotheriids, right whales (balaenids), and grey whales (eschrichtiids).

The main difference between each family of mysticete is in their feeding adaptations and subsequent behaviour. Balaenopterids are the rorquals. These animals, along with the cetotheriids, rely on their throat pleats to gulp large amounts of water while feeding. The throat pleats extend from the mouth to the navel and allow the mouth to expand to a large volume for more efficient capture of the small animals they feed on. Balaenopterids consist of two genera and eight species. Balaenids are the right whales. These animals have very large heads, which can make up as much as 40% of their body mass, and much of the head is the mouth. This allows them to take in large amounts of water into their mouths, letting them feed more effectively. Eschrichtiids have one living member: the grey whale. They are bottom feeders, mainly eating crustaceans and benthic invertebrates. They feed by turning on their sides and taking in water mixed with sediment, which is then expelled through the baleen, leaving their prey trapped inside. This is an efficient method of hunting, in which the whale has no major competitors.

Odontocetes

Odontocetes are known as toothed whales; they have teeth and only one blowhole. They rely on their well-developed sonar to find their way in the water. Toothed whales send out ultrasonic clicks using the melon. Sound waves travel through the water. Upon striking an object in the water, the sound waves bounce back at the whale. These vibrations are received through fatty tissues in the jaw, which is then rerouted into the ear-bone and into the brain where the vibrations are interpreted. All toothed whales are opportunistic, meaning they will eat anything they can fit in their throat because they are unable to chew. These animals rely on their well-developed flippers and tail fin to propel themselves through the water; they swim by moving their fore-flippers and tail fin up and down. Whale ribs loosely articulate with their thoracic vertebrae at the proximal end, but they do not form a rigid rib cage. This adaptation allows the chest to compress during deep dives as opposed to resisting the force of water pressure. Excluding dolphins and porpoises, odontocetes consist of four families: belugas and narwhals (monodontids), sperm whales (physeterids), dwarf and pygmy sperm whales (kogiids), and beaked whales (ziphiids). There are six species, sometimes referred to as "blackfish", that are dolphins commonly misconceived as whales: the killer whale, the melon-headed whale, the pygmy killer whale, the false killer whale, and the two species of pilot whales, all of which are classified under the family Delphinidae (oceanic dolphins).

The differences between families of odontocetes include size, feeding adaptations and distribution. Monodontids consist of two species: the beluga and the narwhal. They both reside in the frigid arctic and both have large amounts of blubber. Belugas, being white, hunt in large pods near the surface and around pack ice, their coloration acting as camouflage. Narwhals, being black, hunt in large pods in the aphotic zone, but their underbelly still remains white to remain camouflaged when something is looking directly up or down at them. They have no dorsal fin to prevent collision with pack ice. Physeterids and Kogiids consist of sperm whales. Sperm whales consist the largest and smallest odontocetes, and spend a large portion of their life hunting squid. *P. macrocephalus* spends most of its life in search of squid in the depths; these animals do not require any degree of light at all, in fact, blind sperm whales have been caught in perfect health. The behaviour of Kogiids remains largely unknown, but, due to their small lungs, they are thought to hunt in the photic zone. Ziphiids consist of 22 species of beaked whale. These vary from size, to coloration, to distribution, but they all share a similar hunting style. They use a suction technique, aided by a pair of grooves on the underside of their head, not unlike the throat pleats on the rorquals, to feed.

Evolution

Basilosaurus skeleton

Whales are descendants of land-dwelling mammals of the artiodactyl order (even-toed ungulates). They are related to the *Indohyus*, an extinct chevrotain-like ungulate, from which they split approximately 48 million years ago. Primitive cetaceans, or archaeocetes, first took to the sea approximately 49 million years ago and became fully aquatic 5–10 million years later. What defines an archaeocete is the presence of anatomical features exclusive to cetaceans, alongside other primitive features not found in modern cetaceans, such as visible legs or asymmetrical teeth. Their features became adapted for living in the marine environment. Major anatomical changes included their hearing set-up that channeled vibrations from the jaw to the earbone (*Ambulocetus* 49 mya), a streamlined body and the growth of flukes on the tail (*Protocetus* 43 mya), the migration of the nostrils toward the top of the cranium (blowholes), and the modification of the forelimbs into flippers (*Basilosaurus* 35 mya), and the shrinking and eventual disappearance of the hind limbs (the first odontocetes and mysticetes 34 mya).

Whale morphology shows a number of examples of convergent evolution, the most obvious being the streamlined fish-like body shape. Other examples include the use of echolocation for hunting in low light conditions — which is the same hearing adaptation used by bats — and, in the rorqual whales, jaw adaptations, similar to those found in pelicans, that enable engulfment feeding.

Today, the closest living relatives of cetaceans are the hippopotamuses; these share a semi-aquatic ancestor that branched off from other artiodactyls some 60 mya. Around 40 mya, a common ancestor between the two branched off into cetacea and anthracotheres; nearly all anthracotheres became extinct at the end of the Pleistocene 2.5 mya, eventually leaving only one surviving lineage – the hippopotamus.

Whales split into two separate parvorders around 34 mya – the baleen whales (Mysticetes) and the toothed whales (Odontocetes).

Biology

Anatomy

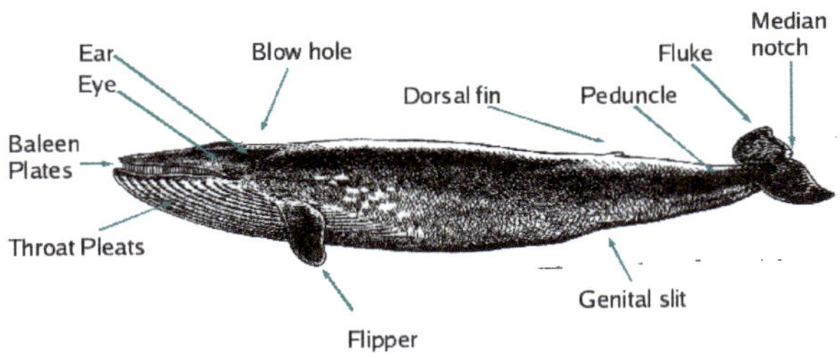

Features of a blue whale

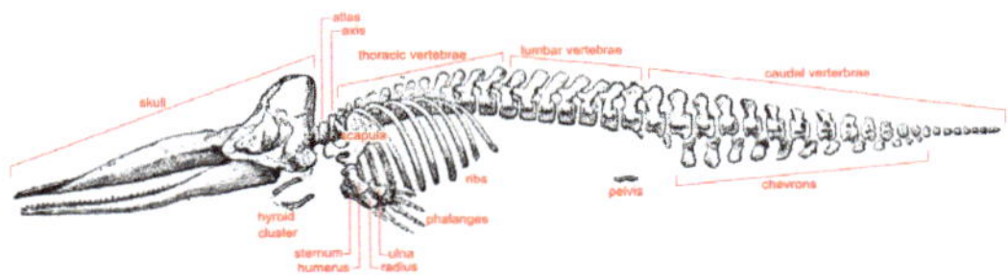

Features of a sperm whale skeleton

Whales have torpedo shaped bodies with non-flexible necks, limbs modified into flippers, non-existent external ear flaps, a large tail fin, and flat heads (with the exception of monodontids and ziphiids). Whale skulls have small eye orbits, long snouts (with the exception of monodontids and ziphiids) and eyes placed on the sides of its head. Whales range in size from the 2.6-metre (8.5 ft) and 135-kilogram (298 lb) dwarf sperm whale to the 34-metre (112 ft) and 190-metric-ton (210-short-ton) blue whale. Overall, they tend to dwarf other cetartiodactyls; the blue whale is the largest creature on earth. Several species have female-biased sexual dimorphism, with the females being larger than the males. One exception is with the sperm whale, which has males larger than the females.

Odontocetes, such as the sperm whale, possess teeth with cementum cells overlying dentine cells. Unlike human teeth, which are composed mostly of enamel on the portion of the tooth outside of the gum, whale teeth have cementum outside the gum. Only in larger whales, where the cementum is worn away on the tip of the tooth, does enamel show. Mysticetes have large whalebone, as opposed to teeth, made of keratin. Mysticetes have two blowholes, whereas Odontocetes contain only one.

Breathing involves expelling stale air from the blowhole, forming an upward, steamy spout, followed by inhaling fresh air into the lungs; a humpback whale's lungs can hold about 5,000 litres of air. Spout shapes differ among species, which facilitates identification.

All whales have a thick layer of blubber. In species that live near the poles, the blubber can be as thick as 11 inches. This blubber can help with buoyancy (which is helpful for a 100-ton whale), protection to some extent as predators would have a hard time getting through a thick layer of fat, and energy for fasting when migrating to the equator; the primary usage for blubber is insulation from the harsh climate. It can constitute as much as 50% of a whale's body weight. Calves are born with only a thin layer of blubber, but some species compensate for this with thick lanugos.

Whales have a two- to three-chambered stomach that is similar in structure to terrestrial carnivores. Mysticetes contain a proventriculus as an extension of the oesophagus; this contains stones that grind up food. They also have fundic and pyloric chambers.

Locomotion

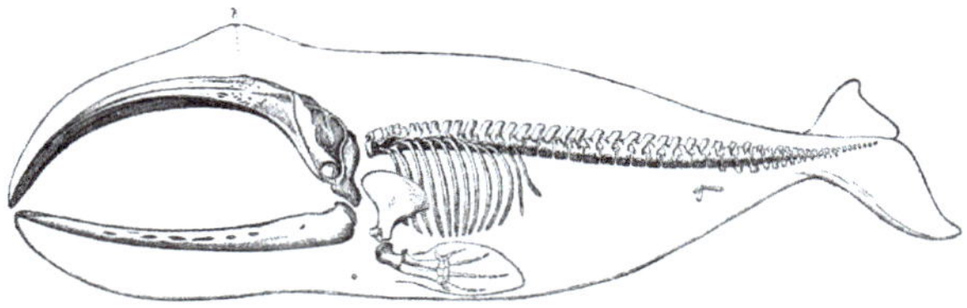

Skeleton of a bowhead whale; notice the hind limb. Richard Lydekker, 1894

Whales have two flippers on the front, and a tail fin. These flippers contain four digits. Although whales do not possess fully developed hind limbs, some, such as the sperm whale and bowhead whale, possess discrete rudimentary appendages, which may contain feet and digits. Whales are fast swimmers in comparison to seals, which typically cruise at 5–15 kn, or 9–28 kilometres per hour (5.6–17.4 mph); the fin whale, in comparison, can travel at speeds up to 47 kilometres per hour (29 mph) and the sperm whale can reach speeds of 35 kilometres per hour (22 mph). The fusing of the neck vertebrae, while increasing stability when swimming at high speeds, decreases flexibility; whales are unable to turn their heads. When swimming, whales rely on their tail fin propel them through the water. Flipper movement is continuous. Whales swim by moving their tail fin and lower body up and down, propelling themselves through vertical movement, while their flippers are mainly used for steering. Some species log out of the water, which may allow then to travel faster. Their skeletal anatomy allows them to be fast swimmers. Most species have a dorsal fin.

Whales are adapted for diving to great depths. In addition to their streamlined bodies, they can slow their heart rate to conserve oxygen; blood is rerouted from tissue tolerant of water pressure to the heart and brain among other organs; haemoglobin and myoglobin store oxygen in body tissue; and they have twice the concentration of myoglobin than haemoglobin. Before going on long dives, many whales exhibit a behaviour known as sounding; they stay close to the surface for a series of short, shallow dives while building their oxygen reserves, and then make a sounding dive.

Senses

The whale ear has specific adaptations to the marine environment. In humans, the middle ear works as an impedance equalizer between the outside air's low impedance and the cochlear fluid's high impedance. In whales, and other marine mammals, there is no great difference between the outer and inner environments. Instead of sound passing through the outer ear to the middle ear, whales receive sound through the throat, from which it passes through a low-impedance fat-filled cavity to the inner ear. The whale ear is acoustically isolated from the skull by air-filled sinus pockets, which allow for greater directional hearing underwater. Odontocetes send out high frequency clicks from an organ known as a melon. This melon consists of fat, and the skull of any such creature containing a melon will have a large depression. The melon size varies between species, the bigger the more dependent they are of it. A beaked whale for example has a small bulge sitting on top of its skull, whereas a sperm whale's head is filled up mainly with the melon.

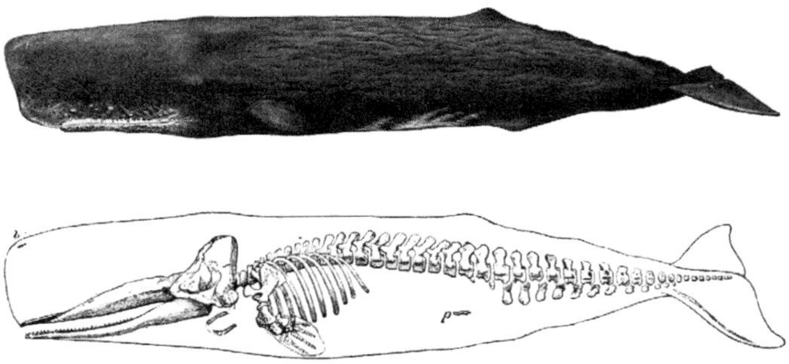

Sperm whale skeleton. Richard Lydekker, 1894.

The whale eye is relatively small for its size, yet they do retain a good degree of eyesight. As well as this, the eyes of a whale are placed on the sides of its head, so their vision consists of two fields, rather than a binocular view like humans have. When belugas surface, their lens and cornea correct the nearsightedness that results from the refraction of light; they contain both rod and cone cells, meaning they can see in both dim and bright light, but they have far more rod cells than they do cone cells. Whales do, however, lack short wavelength sensitive visual pigments in their cone cells indicating a more limited capacity for colour vision than most mammals. Most whales have slightly flattened eyeballs, enlarged pupils (which shrink as they surface to prevent damage), slightly flattened corneas and a tapetum lucidum; these adaptations allow for large amounts of light to pass through the eye and, therefore, a very clear image of the surrounding area. In water, a whale can see around 10.7 metres (35 ft) ahead of itself, but they have a smaller range above water. They also have glands on the eyelids and outer corneal layer that act as protection for the cornea.

The olfactory lobes are absent in toothed whales, suggesting that they have no sense of smell. Some whales, such as the bowhead whale, possess a vomeronasal organ, which does mean that they can "sniff out" krill.

Whales are not thought to have a good sense of taste, as their taste buds are atrophied or missing altogether. However, some toothed whales have preferences between different kinds of fish, indicating some sort of attachment to taste. The presence of the Jacobson's organ indicates that whales can smell food once inside their mouth, which might be similar to the sensation of taste.

Communication

Whale vocalization is likely to serve several purposes. Some species, such as the humpback whale, communicate using melodic sounds, known as whale song. These sounds may be extremely loud, depending on the species. Humpback whales only have been heard making clicks, while toothed whales use sonar that may generate up to 20,000 watts of sound (+73 dBm or +43 dBw) and be heard for many miles.

Captive whales have occasionally been known to mimic human speech. Scientists have suggested this indicates a strong desire on behalf of the whales to communicate with humans, as whales have a very different vocal mechanism, so imitating human speech likely takes considerable effort.

Whales emit two distinct kinds of acoustic signals, which are called whistles and clicks: Clicks are quick broadband burst pulses, used for sonar, although some lower-frequency broadband vocalizations may serve a non-echolocative purpose such as communication; for example, the pulsed calls of belugas. Pulses in a click train are emitted at intervals of ~35–50 milliseconds, and in general these inter-click intervals are slightly greater than the round-trip time of sound to the target. Whistles are narrow-band frequency modulated (FM) signals, used for communicative purposes, such as contact calls.

Intelligence

Whales are known to teach, learn, cooperate, scheme, and grieve. The neocortex of many species of whale is home to elongated spindle neurons that, prior to 2007, were known only in hominids. In humans, these cells are involved in social conduct, emotions, judgement, and theory of mind. Whale spindle neurons are found in areas of the brain that are homologous to where they are found in humans, suggesting that they perform a similar function.

Bubble net feeding

Brain size was previously considered a major indicator of the intelligence of an animal. Since most of the brain is used for maintaining bodily functions, greater ratios of brain to body mass may increase the amount of brain mass available for more complex cognitive tasks. Allometric analysis indicates that mammalian brain size scales at approximately the ⅔ or ¾ exponent of the body mass. Comparison of a particular animal's brain size with the expected brain size based on such allometric analysis provides an encephalisation quotient that can be used as another indication of animal intelligence. Sperm whales have the largest brain mass of any animal on earth, averaging 8,000 cubic centimetres (490 in^3) and 7.8 kilograms (17 lb) in mature males, in comparison to the average human brain which averages 1,450 cubic centimetres (88 in^3) in mature males. The brain to body mass ratio in some odontocetes, such as belugas and narwhals, is second only to humans.

Small whales are known to engage in complex play behaviour, which includes such things as producing stable underwater toroidal air-core vortex rings or "bubble rings". There are two main methods of bubble ring production: rapid puffing of a burst of air into the water and allowing it to rise to the surface, forming a ring, or swimming repeatedly in a circle and then stopping to inject air into the helical vortex currents thus formed. They also appear to enjoy biting the vortex-rings, so that they burst into many separate bubbles and then rise quickly to the surface. Some believe this is a means of communication. Whales are also known to produce bubble-nets for the purpose of foraging.

A southern right whale sailing

Larger whales are also thought, to some degree, to engage in play. The southern right whale, for example, elevates their tail fluke above the water, remaining in the same position for a considerable amount of time. This is known as "sailing". It appears to be a form of play and is most commonly seen off the coast of Argentina and South Africa. Humpback whales, among others, are also known to display this behaviour.

Life Cycle

Whales are fully aquatic creatures, which means that birth and courtship behaviours are very different from terrestrial and semi-aquatic creatures. Since they are unable to go onto land to calve, they deliver the baby with the fetus positioned for tail-first delivery. This prevents the baby from drowning either upon or during delivery. To feed the new-born, whales, being aquatic, must squirt the milk into the mouth of the calf. Being mammals, they have mammary glands used for nursing calves; they are weaned off at about 11 months of age. This milk contains high amounts of fat which is meant to hasten the development of blubber; it contains so much fat that it has the consistency of toothpaste. Females deliver a single calf with gestation lasting about a year, dependency until one to two years, and maturity around seven to ten years, all varying between the species. This mode of reproduction produces few offspring, but increases the survival probability of each one. Females, referred to as "cows", carry the responsibility of childcare as males, referred to as "bulls", play no part in raising calves.

Most mysticetes reside at the poles. So, to prevent the unborn calf from dying of frostbite, they migrate to calving/mating grounds. They will then stay there for a matter of months until the calf has developed enough blubber to survive the bitter temperatures of the poles. Until then, the calves will feed on the mother's fatty milk. With the exception of the humpback whale, it is largely unknown when whales migrate. Most will travel from the Arctic or Antarctic into the tropics to mate, calve, and raise during the winter and spring; they will migrate back to the poles in the warmer summer months so the calf can continue growing while the mother can continue eating, as they fast in the breeding grounds. One exception to this is the southern right whale, which migrates to Patagonia and western New Zealand to calve; both are well out of the tropic zone.

Sleep

Unlike most animals, whales are conscious breathers. All mammals sleep, but whales cannot afford to become unconscious for long because they may drown. While knowledge of sleep in wild cetaceans is limited, toothed cetaceans in captivity have been recorded to sleep with one side of their brain at a time, so that they may swim, breathe consciously, and avoid both predators and social contact during their period of rest.

A 2008 study found that sperm whales sleep in vertical postures just under the surface in passive shallow 'drift-dives', generally during the day, during which whales do not respond to passing vessels unless they are in contact, leading to the suggestion that whales possibly sleep during such dives.

Ecology

Foraging and Predation

Polar bear with the remains of a beluga

All whales are carnivorous and predatory. Odontocetes, as a whole, mostly feed on fish and cephalopods, and then followed by crustaceans and bivalves. All species are generalist and opportunistic feeders. Mysticetes, as a whole, mostly feed on krill and plankton, followed by crustaceans and other invertebrates. A few are specialists. Examples include the blue whale, which eats almost exclusively krill, the minke whale, which eats mainly schooling fish, the sperm whale, which specialize on squid, and the grey whale which feed on bottom-dwelling invertebrates. The elaborate baleen "teeth" of filter-feeding species, mysticetes, allow them to remove water before they swallow their planktonic food by using the teeth as a sieve. Usually whales hunt solitarily, but they do sometimes hunt cooperatively in small groups. The former behaviour is typical when hunting non-schooling fish, slow-moving or immobile invertebrates or endothermic prey. When large amounts of prey are available, whales such as certain mysticetes hunt cooperatively in small groups. Some cetaceans may forage with other kinds of animals, such as other species of whales or certain species of pinnipeds.

Large whales, such as mysticetes, are not usually subject to predation, but smaller whales, such as monodontids or ziphiids, are. These species are preyed on by the killer whale or orca. To subdue and kill whales, orcas continuously ram them with their heads; this can sometimes kill bowhead whales, or severely injure them. Other times they corral the narwhals or belugas before striking. They are typically hunted by groups of 10 or fewer orcas, but they are seldom attacked by an individual. Calves are more commonly taken by orcas, but adults can be targeted as well.

These small whales are also targeted by terrestrial and pagophilic predators. The polar bear is well adapted for hunting Arctic whales and calves. Bears are known to use sit-and-wait tactics as well as active stalking and pursuit of prey on ice or water. Whales lessen the chance of predation by gathering in groups. This however means less room around the breathing hole as the ice slowly closes the gap. When out at sea, whales dive out of the reach of surface-hunting orcas. Polar bear attacks on belugas and narwhals are usually successful in winter, but rarely inflict any damage in summer.

Whale Pump

A 2010 study considered whales to be a positive influence to the productivity of ocean fisheries, in what has been termed a "whale pump." Whales carry nutrients such as nitrogen from the depths back to the surface. This functions as an upward biological pump, reversing an earlier presumption that whales accelerate the loss of nutrients to the bottom. This nitrogen input in the Gulf of Maine is "more than the input of all rivers combined" emptying into the gulf, some 23,000 metric tons (25,000 short tons) each year. Whales defecate at the ocean's surface; their excrement is important for fisheries because it is rich in iron and nitrogen. The whale faeces are liquid and instead of sinking, they stay at the surface where phytoplankton feed off it.

Whale Fall

Upon death, whale carcasses fall to the deep ocean and provide a substantial habitat for marine life. Evidence of whale falls in present-day and fossil records shows that deep sea whale falls support a rich assemblage of creatures, with a global diversity of 407 species, comparable to other neritic biodiversity hotspots, such as cold seeps and hydrothermal vents.

Deterioration of whale carcasses happens though a series of three stages. Initially, moving organisms such as sharks and hagfish, scavenge the soft tissues at a rapid rate over a period of months, and as long as two years. This is followed by the colonization of bones and surrounding sediments (which contain organic matter) by enrichment opportunists, such as crustaceans and polychaetes, throughout a period of years. Finally, sulfophilic bacteria reduce the bones releasing hydrogen sulfide enabling the growth of chemoautotrophic organisms, which in turn, support other organisms such as mussels, clams, limpets, and sea snails. This stage may last for decades and supports a rich assemblage of species, averaging 185 species per site.

Interaction with Humans

Whaling

Whale Fishing: Woodcut by Thevet, Paris, 1574

Dutch whalers near Spitsbergen, their most successful port. *Abraham Storck, 1690*

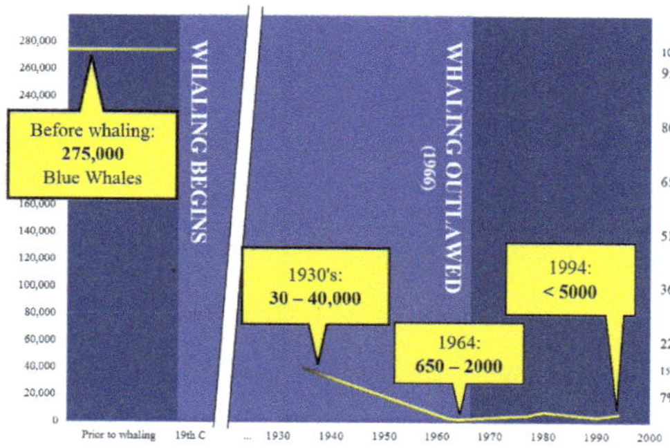

World population graph of blue whales

Whaling by humans has existed since the Stone Age. Ancient whalers used harpoons to spear the bigger animals from boats out at sea. People from Norway and Japan started hunting whales around 2000 B.C. Whales are typically hunted for their meat and blubber by aboriginal groups; they used baleen for baskets or roofing, and made tools and masks out of bones. The Inuit hunted whales in the Arctic Ocean. The Basques started whaling as early as the 11th century, sailing as far as Newfoundland in the 16th century in search of right whales. 18th- and 19th-century whalers hunted whales mainly for their oil, which was used as lamp fuel and a lubricant, baleen or whalebone, which was used for items such as corsets and skirt hoops, and ambergris, which was used as a fixative for perfumes. The most successful whaling nations at this time were the Netherlands, Japan, and the United States.

Commercial whaling was historically important as an industry well throughout the 17th, 18th and 19th centuries. Whaling was at that time a sizeable European industry with ships from Britain, France, Spain, Denmark, the Netherlands and Germany, sometimes collaborating to hunt whales in the Arctic, sometimes in competition leading even to war. By the early 1790s, whalers, namely the Americans and Australians, focused efforts in the South Pacific where they mainly hunted sperm whales and right whales, with catches of up to 39,000 right whales by Americans alone. By 1853, US profits reached US$11,000,000 (UK£6.5m), equivalent to US$348,000,000 (UK£230m) today, the most profitable year for the American whaling industry. Commonly exploited species included North Atlantic right whales, sperm whales, which were mainly hunted by Americans, bowhead whales, which were mainly hunted by the Dutch, common minke whales, blue whales, and grey whales. The scale of whale harvesting decreased substantially after 1982 when the International Whaling Commission (IWC) placed a moratorium which set a catch limit for each country, excluding aboriginal groups until 2004.

Current whaling nations are Norway, Iceland, and Japan, despite their joining to the IWC, as well as the aboriginal communities of Siberia, Alaska, and northern Canada. Subsistence hunters typically use whale products for themselves and depend on them for survival. National and international authorities have given special treatment to aboriginal hunters since their methods of hunting are seen as less destructive and wasteful. This distinction is being questioned as these aboriginal groups are using more modern weaponry and mechanized transport to hunt with, and are selling whale products in the marketplace. Some anthropologists argue that the term "subsistence" should also apply to these cash-based exchanges as long as they take place within local production and consumption. In 1946, the IWC placed a moratorium, limiting the annual whale catch. Since then, yearly profits for these "subsistence" hunters have been close to US$31 million (UK£20m) per year.

Other Threats

Whales can also be threatened by humans more indirectly. They are unintentionally caught in fishing nets by commercial fisheries as bycatch and accidentally swallow fishing hooks. Gillnetting and Seine netting is a significant cause of mortality in whales and other marine mammals. Species commonly entangled include beaked whales. Whales are also affected by marine pollution. High levels of organic chemicals accumulate in these animals since they are high in the food chain. They have large reserves of blubber, more so for toothed whales as they are higher up the food chain than baleen whales. Lactating mothers can pass the toxins on to their young. These pollutants can cause gastrointestinal cancers and greater vulnerability to infectious diseases. They can also be poisoned by swallowing litter, such as plastic bags. Environmentalists speculate that advanced naval sonar endangers some whales. Some scientists suggest that sonar may trigger whale beachings, and they point to signs that such whales have experienced decompression sickness.

Conservation

World map showing International Whaling Commission (IWC) members in blue

Whaling decreased substantially after 1946 when, in response to the steep decline in whale populations, the International Whaling Commission placed a moratorium which set a catch limit for each country; this excluded aboriginal groups up until 2004. As of 2015, aboriginal communities are allowed to take 280 bowhead whales off of Alaska and two from the western coast of Greenland, 620 grey whales off Washington state, three common minke whales off of the eastern coast of Greenland and 178 on their western coast, 10 fin whales from the west coast of Greenland, nine humpback whales from the west coast of Greenland and 20 off of St. Vincent and the Grenadines each year. Several species that were commercially exploited have rebounded in numbers; for example, grey whales may be as numerous as they were prior to harvesting, but the North Atlantic population is functionally extinct. Conversely, the North Atlantic right whale was extirpated from much of its former range, which stretched across the North Atlantic, and only remains in small fragments along the coast of Canada, Greenland, and is considered functionally extinct along the European coastline.

The IWC has designated two whale sanctuaries: the Southern Ocean Whale Sanctuary, and the Indian Ocean Whale Sanctuary. The Southern Ocean whale sanctuary spans 30,560,860 square kilometres (11,799,610 sq mi) and envelopes Antarctica. The Indian Ocean whale sanctuary takes up all of the Indian Ocean south of 55°S. The IWC is a voluntary organization, with no treaty. Any nation may leave as they wish; the IWC cannot enforce any law it makes.

As of 2013, the International Union for Conservation of Nature (IUCN) recognized 86 cetacean species, 40 of which are considered whales. Six are considered at risk, as they are ranked Critically Endangered (the North Atlantic right whale), "Endangered" (blue whale, fin whale, North Pacific right whale, and sei whale), and "Vulnerable" (sperm whale). Twenty-one species have a "Data Deficient" ranking. Species that live in polar habitats are vulnerable to the effects of recent and ongoing climate change, particularly the time when pack ice forms and melts.

Whale Watching

Whale watching off Bar Harbour, Maine

An estimated 13 million people went whale watching globally in 2008, in all oceans except the Arctic. Rules and codes of conduct have been created to minimize harassment of the whales. Iceland, Japan and Norway have both whaling and whale watching industries. Whale watching lobbyists are concerned that the most inquisitive whales, which approach boats closely and provide much of the entertainment on whale-watching trips, will be the first to be taken if whaling is resumed in the same areas. Whale watching generated US$2.1 billion (UK£1.4 billion) per annum in tourism revenue worldwide, employing around 13,000 workers. In contrast, the whaling industry, with the moratorium in place, generates US$31 million (UK£20 million) per year. The size and rapid growth of the industry has led to complex and continuing debates with the whaling industry about the best use of whales as a natural resource.

In Myth, Literature and Art

As marine creatures that reside in either the depths or the poles, humans knew very little about whales over the course of history; many feared or revered them. The Nords and various arctic tribes revered the whale as they were important pieces of their lives. In Inuit creation myths, when 'Big Raven', a deity in human form, found a stranded whale, he was told by the Great Spirit where to find special mushrooms that would give him the strength to drag the whale back to the sea and thus, return order to the world. In an Icelandic legend, a man threw a stone at a fin whale and hit the blowhole, causing the whale to burst. The man was told not to go to sea for twenty years, but during the nineteenth year he went fishing and a whale came and killed him.

Engraving by William van der Gouwen depicting a stranded sperm whale being butchered on the Dutch coast, 1598

Whales played a major part in shaping the art forms of many coastal civilizations, such as the Norse, with some dating to the Stone Age. Petroglyphs off a cliff face in Bangudae, South Korea show 300 depictions of various animals, a third of which are whales. Some show particular detail in which there are throat pleats, typical of rorquals. These petroglyphs show these people, of around 7,000 to 3,500 B.C.E. in South Korea, had a very high dependency on whales.

Whalers off Twofold Bay, New South Wales. Watercolour by Oswald Brierly, 1867

The Pacific Islanders and Australian Aborigines viewed whales as bringers of good and joy. One exception is French Polynesia, where, in many parts, cetaceans are met with great brutality.

In Vietnam and Ghana, among other places, whales hold a sense of divinity. They are so respected in their cultures that they occasionally hold funerals for beached whales, a throwback to Vietnam's ancient sea-based Austro-Asiatic culture. The god of the seas, according to Chinese folklore, was a large whale with human limbs.

Whales have also played a role in sacred texts such as the Bible. It mentions whales in Genesis 1:21, Job 7:12, and Ezekiel 32:2. The "leviathan" described at length in Job 41:1-34 is generally understood to refer to a whale. The "sea monsters" in Lamentations 4:3 have been taken by some to refer to marine mammals, in particular whales, although most modern versions use the word "jackals" instead. The story of Jonah being swallowed by a great fish is told both in the Qur'an and in the Bible. A medieval column capital sculpture depicting this was made in the 12th century in the abbey church in Mozac, France. The Old Testament contains the Book of Jonah and in the New Testament, Jesus mentions this story in Matthew 12:40.

In 1585, Alessandro Farnese, 1585, and Francois, Duke of Anjou, 1582, were greeted on his ceremonial entry into the port city of Antwerp by floats including "Neptune and the Whale", indicating at least the city's dependence on the sea for its wealth.

Whales continue to be prevalent in modern literature. For example, Herman Melville's *Moby Dick* features a "great white whale" as the main antagonist for Ahab, who eventually is killed by it. The whale is an albino sperm whale, considered by Melville to be the largest type of whale, and is partly based on the historically attested bull whale Mocha Dick. Rudyard Kipling's *Just So Stories* includes the story of "How the Whale got in his Throat". Niki Caro's film the *Whale Rider* has a Māori girl ride a whale in her journey to be a suitable heir to the chieftain-ship. Walt Disney's film *Pinocchio* features a giant whale named Monstro as the final antagonist. Alan Hovhaness' orchestra *And God Created Great Whales* including the recorded sounds of humpback and bowhead whales. Léo Ferré's song "Il n'y a plus rien" is an example of biomusic that begins and ends with recorded whale songs mixed with a symphonic orchestra and his voice.

In Captivity

Beluga whales and trainers in an aquarium

Belugas were the first whales to be kept in captivity. Other species were too rare, too shy, or too big. The first beluga was shown at Barnum's Museum in New York City in 1861. For most of the 20th century, Canada was the predominant source of wild belugas. They were taken from the St. Lawrence River estuary until the late 1960s, after which they were predominantly taken from the Churchill River estuary until capture was banned in 1992. Russia has become the largest provider since it had been banned in Canada. Belugas are caught in the Amur River delta and their eastern coast, and then are either transported domestically to aquariums or dolphinariums in Moscow, St. Petersburg, and Sochi, or exported to other countries, such as Canada. Most captive belugas are caught in the wild, since captive-breeding programs are not very successful.

As of 2006, 30 belugas were in Canada and 28 in the United States, and 42 deaths in captivity had been reported up to that time. A single specimen can reportedly fetch up to US$100,000 (UK£64,160) on the market. The beluga's popularity is due to its unique colour and its facial expressions. The latter is possible because while most cetacean "smiles" are fixed, the extra movement afforded by the beluga's unfused cervical vertebrae allows a greater range of apparent expression.

Between 1960 and 1992, the Navy carried out a program that included the study of marine mammals' abilities with sonar, with the objective of improving the detection of underwater objects. A large number of belugas were used from 1975 on, the first being dolphins. The program also included training them to carry equipment and material to divers working underwater by holding cameras in their mouths to locate lost objects, survey ships and submarines, and underwater monitoring. A similar program was used by the Russian Navy during the Cold War, in which belugas were also trained for antimining operations in the Arctic.

Aquariums have tried housing other species of whales in captivity. The success of belugas turned attention to maintaining their relative, the narwhal, in captivity. However, in repeated attempts in the 1960s and 1970s, all narwhals kept in captivity died within months. A pair of pygmy right whales were retained in an enclosed area (with nets); they were eventually released in South Africa. There was one attempt to keep a stranded Sowerby's beaked whale calf in captivity; the calf rammed into the tank wall, breaking its rostrum, which resulted in death. It was thought that Sowerby's beaked whale evolved to swim fast in a straight line, and a 30 metres (98 ft) was not big enough. There have been attempts to keep baleen whales in captivity. There were three attempts to keep grey whales in captivity. Gigi was a grey whale calf that died in transport. Gigi II was another grey whale calf that was captured in the Ojo de Liebre Lagoon, and was transported to SeaWorld. The 680 kilograms (1,500 lb) calf was a popular attraction, and behaved normally, despite being separated from his mother. A year later, the 8,000-kilogram (18,000 lb) whale grew too big to keep in captivity and was released; it was the first of two grey whales, the other being another grey whale calf named JJ, to successfully be kept in captivity. There were three attempts to keep minke whales in captivity in Japan. They were kept in a tidal pool with a sea-gate at the Izu Mito Sea Paradise. Another, unsuccessful, attempt was made by the U.S. One stranded humpback whale calf was kept in captivity for rehabilitation, but died days later.

References

- Romero, S.; and Green, S.M. (2005). The end of regressive evolution: examining and interpreting the evidence from cave fishes. Journal of Fish Biology 67(1): 3-32
- Romero, Aldemaro, editor (2001). The Biology of Hypogean Fishes. Developments in Environmental Biology of Fishes. ISBN 978-1402000768
- Owen, J. (11 September 2015). "How This Cave-Dwelling Fish Lost Its Eyes to Evolution". National Geographic. Retrieved 14 May 2017
- Behrmann-Godel, J.; A.W. Nolte; J. Kreiselmaier; R. Berka; J. Freyhof (2017). "The first European cave fish". Current Biology. 27 (7): R257–R258. doi:10.1016/j.cub.2017.02.048
- Jeffery; Strickler; and Yamamoto (2003). "To See or Not to See: Evolution of Eye Degeneration in Mexican Blind Cavefish". Integr Comp Biol. 43 (4): 531–541. doi:10.1093/icb/43.4.531
- Andy Coghlan (3 April 2017). "First ever cavefish discovered in Europe evolved super-fast". New Scientist. Retrieved 4 April 2017
- Zhang, C. & Zhao, Y.-H. (2016). Species Diversity and Distribution of Inland Fishes in China. Science Press. p. 296. ISBN 9787030472106
- Tencatt; and Bichuette (2017). "Aspidoras mephisto, new species: The first troglobitic Callichthyidae (Teleostei: Siluriformes) from South America". PLoS ONE. 12 (3): e0171309. doi:10.1371/journal.pone.0171309
- Rice, D. W. (1 January 1998). Marine Mammals of the World: Systematics and Distribution (PDF). Society for Marine Mammalogy. pp. 92–95. ISBN 978-1-891276-03-3. OCLC 40622084

- University of Texas at Austin (17 June 2016). "Rare, blind catfish never before found in US discovered in national park cave in Texas". ScienceDaily. Retrieved 13 May 2017
- Reidenberg, Joy S. (2007). "Anatomical adaptations of aquatic mammals". The Anatomical Record. 290 (6): 507–513. PMID 17516440. doi:10.1002/ar.20541
- Espinasa, L.; and W.R. Jeffery (2003). "A troglomorphic sculpin (Pisces: Cottidae) population: geography, morphology and conservation status". Journal of Cave and Karst Studies. 65 (2): 93–100
- Smith, Brian D.; Perrin, William F. (1997). "Distribution, Mortality, Diet and Conservation of Irrawaddy Dolphins (Orcaella Brevirostris) In Lao PDR". Asian Marine Biology. 14: 41–48. ISBN 978-962-209-462-8
- Shaji, C.P. (2011). "Indoreonectes evezardi". IUCN Red List of Threatened Species. Version 2017.1. International Union for Conservation of Nature. Retrieved 13 May 2017
- Hamilton, H.; S. Caballero; A. G. Collins & R. L. Brownell Jr. (2001). "Evolution of river dolphins". Proceedings of the Royal Society B. 268 (1466): 549–556. PMC 1088639 . PMID 11296868. doi:10.1098/rspb.2000.1385
- Edwards, Holly H.; Schnell, Gary D. (2001). "Body Length, Swimming Speed, Dive Duration, and Coloration of the Dolphin Sotalia fluviatilis (Tucuxi) in Nicaragua" (PDF). Caribbean Journal of Science. 37: 271–298
- Thomas, Jeanette A.; Kastelein, Ronald A., eds. (2002). Sensory Abilities of Cetaceans: Laboratory and Field M. A. Cravalho (1999). "Shameless creatures: An ethnozoology of the Amazon river dolphin". Ethnology. 38 (1): 47–58. doi:10.2307/3774086
- Lodi, L.; Barreto, A. (1998). "Legal actions taken in Brazil for the conservation of cetaceans". Journal of International Wildlife Law and Policy. I (3): 403–411. doi:10.1080/13880299809353910
- Kielan-Jaworowska, Zofia; Cifelli, Richard L.; Luo, Zhe-Xi (2004). Mammals from the Age of Dinosaurs: Origins, Evolution, and Structure. New York: Columbia University Press. pp. 441–462. ISBN 978-0-231-50927-5
- Gibbons, Johnny (September 1, 2015). "Fossil Specimen Reveals a New Species of Ancient River Dolphin". Smithsonian Science News. Retrieved September 1, 2015
- Jefferson, T. A.; Webber, M. A.; Pitman, R. L. (2009). Marine Mammals of the World A Comprehensive Guide to their Identification (1 ed.). London: Academic Press. pp. 7–16. ISBN 978-0-12-383853-7

4

Marine Invertebrates: An Integrated Study

Marine invertebrates are animals without a vertebral column that are found within marine habitats. Some of the marine invertebrates are annelid, brachiopod, bryozoa and chaetognatha. The aspects elucidated in this chapter are of vital importance, and provide a better understanding of marine invertebrates.

Marine Invertebrates

Ernst Haeckel's 96th plate, showing some marine invertebrates.
Marine invertebrates have a large variety of body plans, which are currently categorised into over 30 phyla.

Marine invertebrates are the invertebrates that live in marine habitats. Invertebrate is a blanket term that includes all animals apart from the vertebrate members of the chordate phylum. Invertebrates lack a vertebral column, and some have evolved a shell or a hard exoskeleton. As on land and in the air, marine invertebrates have a large variety of body plans, and have been categorised into over 30 phyla. They make up most of the macroscopic life in the oceans.

Evolution

The earliest animals were marine invertebrates, that is, vertebrates came later. Animals are multicellular eukaryotes, and are distinguished from plants, algae, and fungi by lacking cell walls.

Marine invertebrates are animals that inhabit a marine environment apart from the vertebrate members of the chordate phylum; invertebrates lack a vertebral column. Some have evolved a shell or a hard exoskeleton.

Kimberella, an early mollusc important for understanding the Cambrian explosion. Invertebrates are grouped into different phyla (body plans).

Opabinia an extinct, stem group arthropod that appeared in the Middle Cambrian

The earliest widely accepted animal fossils are the rather modern-looking cnidarians (the group that includes jellyfish, sea anemones and *Hydra*), possibly from around 580 Ma The Ediacara biota, which flourished for the last 40 million years before the start of the Cambrian, were the first animals more than a very few centimetres long. Many were flat and had a "quilted" appearance, and seemed so strange that there was a proposal to classify them as a separate kingdom, Vendozoa. Others, however, have been interpreted as early molluscs (*Kimberella*), echinoderms (*Arkarua*), and arthropods (*Spriggina, Parvancorina*). There is still debate about the classification of these specimens, mainly because the diagnostic features which allow taxonomists to classify more recent organisms, such as similarities to living organisms, are generally absent in the Ediacarans. However, there seems little doubt that *Kimberella* was at least a triploblastic bilaterian animal, in other words, an animal significantly more complex than the cnidarians.

The small shelly fauna are a very mixed collection of fossils found between the Late Ediacaran and Middle Cambrian periods. The earliest, *Cloudina*, shows signs of successful defense against predation and may indicate the start of an evolutionary arms race. Some tiny Early Cambrian shells almost certainly belonged to molluscs, while the owners of some "armor plates," *Halkieria* and *Microdictyon*, were eventually identified when more complete specimens were found in Cambrian lagerstätten that preserved soft-bodied animals.

In the 1970s there was already a debate about whether the emergence of the modern phyla was "explosive" or gradual but hidden by the shortage of Precambrian animal fossils. A re-analysis of fossils from the Burgess Shale lagerstätte increased interest in the issue when it revealed animals, such as *Opabinia*, which did not fit into any known phylum. At the time these were interpreted as evidence that the modern phyla had evolved very rapidly in the Cambrian explosion and that the Burgess Shale's "weird wonders" showed that the Early Cambrian was a uniquely experimental period of animal evolution. Later discoveries of similar animals and the development of new theoretical approaches led to the conclusion that many of the "weird wonders" were evolutionary "aunts" or "cousins" of modern groups—for example that *Opabinia* was a member of the lobopods, a group which includes the ancestors of the arthropods, and that it may have been closely related to the modern tardigrades. Nevertheless, there is still much debate about whether the Cambrian explosion was really explosive and, if so, how and why it happened and why it appears unique in the history of animals.

Classification

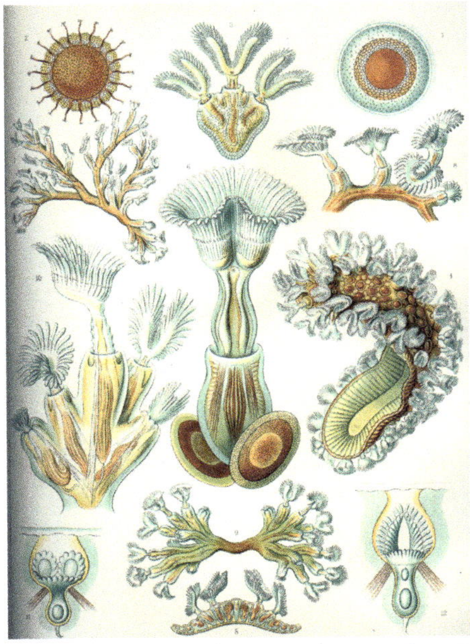

Bryozoa, from Ernst Haeckel's *Kunstformen der Natur*, 1904

Invertebrates are grouped into different phyla. Informally phyla can be thought of as a way of grouping organisms according to their body plan. A body plan refers to a blueprint which describes the shape or morphology of an organism, such as its symmetry, segmentation and the disposition of its appendages. The idea of body plans originated with vertebrates, which were grouped into one phylum. But the vertebrate body plan is only one of many, and invertebrates consist of many phyla or body plans. The history of the discovery of body plans can be seen as a movement from a worldview centred on vertebrates, to seeing the vertebrates as one body plan among many. Among the pioneering zoologists, Linnaeus identified two body plans outside the vertebrates; Cuvier identified three; and Haeckel had four, as well as the Protista with eight more, for a total of twelve. For comparison, the number of phyla recognised by modern zoologists has risen to 35.

Historically body plans were thought of as having evolved in rapidly during the Cambrian explosion, but a more nuanced understanding of animal evolution suggests a gradual development of body plans throughout the early Palaeozoic and beyond. More generally a phylum can be defined in two ways: as described above, as a group of organisms with a certain degree of morphological or developmental similarity (the phenetic definition), or a group of organisms with a certain degree of evolutionary relatedness (the phylogenetic definition).

As on land and in the air, invertebrates make up a great majority of all macroscopic life, as the vertebrates makes up a subphylum of one of over 30 known animal phyla, making the term almost meaningless for taxonomic purpose. Invertebrate sea life includes the following groups, some of which are phyla:

The 49th plate from Ernst Haeckel's *Kunstformen der Natur*, 1904, showing various sea anemones classified as Actiniae, in the Cnidaria phylum

"A variety of marine worms": plate from *Das Meer* by M.J. Schleiden (1804–1881)

- Acoela, among the most primitive bilateral animals;
- Annelida, (polychaetes and sea leeches);

- Brachiopoda, marine animals that have hard "valves" (shells) on the upper and lower surfaces ;
- Bryozoa, also known as moss animals or sea mats;
- Chaetognatha, commonly known as arrow worms, are a phylum of predatory marine worms that are a major component of plankton;
- Cephalochordata represented in the modern oceans by the lancelets (also known as Amphioxus);
- Cnidaria, such as jellyfish, sea anemones, and corals;
- Crustacea, including lobsters, crabs, shrimp, crayfish, barnacles, hermit crabs, mantis shrimps, and copepods;
- Ctenophora, also known as comb jellies, the largest animals that swim by means of cilia;
- Echinodermata, including sea stars, brittle stars, sea urchins, sand dollars, sea cucumbers, crinoids, and sea daisies;
- Echiura, also known as spoon worms;
- Gnathostomulids, slender to thread-like worms, with a transparent body that inhabit sand and mud beneath shallow coastal waters;
- Gastrotricha, often called hairy backs, found mostly interstitially in between sediment particles;
- Hemichordata, includes acorn worms, solitary worm-shaped organisms;
- Kamptozoa, goblet-shaped sessile aquatic animals, with relatively long stalks and a "crown" of solid tentacles, also called Entoprocta;
- Kinorhyncha, segmented, limbless animals, widespread in mud or sand at all depths, also called mud dragons;
- Loricifera, very small to microscopic marine sediment-dwelling animals only discovered in 1983;
- Merostomata; also known as horseshoe crabs;
- Mollusca, including shellfish, squid, octopus, whelks, *Nautilus*, cuttlefish, nudibranchs, scallops, sea snails, Aplacophora, Caudofoveata, Monoplacophora, Polyplacophora, and Scaphopoda;
- Myzostomida, a taxonomic group of small marine worms which are parasitic on crinoids or "sea lilies";
- Nemertinea, also known as "ribbon worms" or "proboscis worms";
- Orthonectida, a small phylum of poorly known parasites of marine invertebrates that are among the simplest of multi-cellular organisms;

- Phoronida, a phylum of marine animals that filter-feed with a lophophore (a "crown" of tentacles), and build upright tubes of chitin to support and protect their soft bodies;
- Placozoa, small, flattened, multicellular animals around 1 millimetre across and the simplest in structure. They have no regular outline, although the lower surface is somewhat concave, and the upper surface is always flattened;
- Porifera (sponges), multicellular organisms that have bodies full of pores and channels allowing water to circulate through them;
- Priapulida, or penis worms, are a phylum of marine worms that live marine mud. They are named for their extensible spiny proboscis, which, in some species, may have a shape like that of a human penis;
- Pycnogonida, also called sea spiders, are unrelated to spiders, or even to arachnids which they resemble;
- Sipunculida, also called peanut worms, is a group containing 144–320 species (estimates vary) of bilaterally symmetrical, unsegmented marine worms;
- Tunicata, also known as sea squirts or sea pork, are filter feeders attached to rocks or similarly suitable surfaces on the ocean floor;
- Some flatworms of the classes Turbellaria and Monogenea;
- Xenoturbella, a genus of bilaterian animals that contains only two marine worm-like species;
- Xiphosura, includes a large number of extinct lineages and only four recent species in the family Limulidae, which include the horseshoe crabs.

Arthropods total about 1,113,000 described extant species, molluscs about 85,000 and chordates about 52,000.

Marine Sponges

Sponges have no nervous, digestive or circulatory system

Sponges are animals of the phylum Porifera (Modern Latin for *bearing pores*). They are multicellular organisms that have bodies full of pores and channels allowing water to circulate through them, consisting of jelly-like mesohyl sandwiched between two thin layers of cells. They have unspecialized cells that can transform into other types and that often migrate between the main cell layers and the mesohyl in the process. Sponges do not have nervous, digestive or circulatory systems. Instead, most rely on maintaining a constant water flow through their bodies to obtain food and oxygen and to remove wastes.

Sponges are similar to other animals in that they are multicellular, heterotrophic, lack cell walls and produce sperm cells. Unlike other animals, they lack true tissues and organs, and have no body symmetry. The shapes of their bodies are adapted for maximal efficiency of water flow through the central cavity, where it deposits nutrients, and leaves through a hole called the osculum. Many sponges have internal skeletons of spongin and/or spicules of calcium carbonate or silicon dioxide. All sponges are sessile aquatic animals. Although there are freshwater species, the great majority are marine (salt water) species, ranging from tidal zones to depths exceeding 8,800 m (5.5 mi).

While most of the approximately 5,000–10,000 known species feed on bacteria and other food particles in the water, some host photosynthesizing micro-organisms as endosymbionts and these alliances often produce more food and oxygen than they consume. A few species of sponge that live in food-poor environments have become carnivores that prey mainly on small crustaceans.

Sponge biodiversity. There are four sponge species in this photo.

Venus' flower basket at a depth of 2572 meters

Barrel sponge

Linnaeus mistakenly identified sponges as plants in the order Algae. For a long time thereafter sponges were assigned to a separate subkingdom, Parazoa (meaning *beside the animals*). They are now classified as a paraphyletic phylum from which the higher animals have evolved.

Marine Cnidarians

Cnidarians (Greek for *nettle*) are distinguished by the presence of stinging cells, specialized cells that they use mainly for capturing prey. Cnidarians include corals, sea anemones, jellyfish and hydrozoans. They form a phylum containing over 10,000 species of animals found exclusively in aquatic (mainly marine) environments. Their bodies consist of mesoglea, a non-living jelly-like substance, sandwiched between two layers of epithelium that are mostly one cell thick. They have two basic body forms: swimming medusae and sessile polyps, both of which are radially symmetrical with mouths surrounded by tentacles that bear cnidocytes. Both forms have a single orifice and body cavity that are used for digestion and respiration.

Cnidarians are the simplest animals with cells organised into tissues.
Yet the starlet sea anemone contains the same genes as those that form the vertebrate head.

Fossil cnidarians have been found in rocks formed about 580 million years ago. Fossils of cnidarians that do not build mineralized structures are rare. Scientists currently think cnidarians, ctenophores and bilaterians are more closely related to calcareous sponges than these are to other sponges, and that anthozoans are the evolutionary "aunts" or "sisters" of other cnidarians, and the most closely related to bilaterians.

Cnidarians are the simplest animals in which the cells are organised into tissues. The starlet sea anemone is used as a model organism in research. It is easy to care for in the laboratory and a protocol has been developed which can yield large numbers of embryos on a daily basis. There is a remarkable degree of similarity in the gene sequence conservation and complexity between the sea anemone and vertebrates. In particular, genes concerned in the formation of the head in vertebrates are also present in the anemone.

Sea anemones are common in tide-pools

Their tentacles sting and paralyse small fish

Close up of polyps on the surface of a coral, waving their tentacles.

Marine Worms

Worms (Old English for *serpent*) typically have long cylindrical tube-like bodies and no limbs. Marine worms vary in size from microscopic to over 1 metre (3.3 ft) in length for some marine polychaete worms (bristle worms) and up to 58 metres (190 ft) for the marine nemertean worm (bootlace worm). Some marine worms occupy a small variety of parasitic niches, living inside the bodies of other animals, while others live more freely in the marine environment or by burrowing underground.

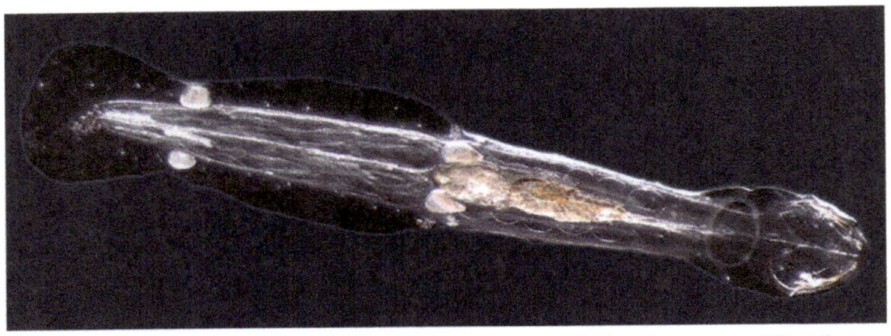

Arrow worms are predatory components of plankton worldwide.

Different groups of marine worms are related only distantly, so they are found in several different phyla such as the Annelida (segmented worms), Chaetognatha (arrow worms), Hemichordata, and Phoronida (horseshoe worms). Many of these worms have specialized tentacles used for exchanging oxygen and carbon dioxide and also may be used for reproduction. Some marine worms are tube worms, such as the giant tube worm which lives in waters near underwater volcanoes and can withstand temperatures up to 90 degrees Celsius.

Platyhelminthes (flatworms) form another worm phylum which includes a class Cestoda of parasitic tapeworms. The marine tapeworm *Polygonoporus giganticus*, found in the gut of sperm whales, can grow to over 30 m (100 ft).

Nematodes (roundworms) constitute a further worm phylum with tubular digestive systems and an opening at both ends. Over 25,000 nematode species have been described, of which more than half are parasitic. It has been estimated another million remain undescribed. They are ubiquitous in marine, freshwater and terrestrial environments, where they often outnumber other animals in both individual and species counts. They are found in every part of the earth's lithosphere, from the top of mountains to the bottom of oceanic trenches. By count they represent 90% of all animals on the ocean floor. Their numerical dominance, often exceeding a million individuals per square meter and accounting for about 80% of all individual animals on earth, their diversity of life cycles, and their presence at various trophic levels point at an important role in many ecosystems.

Giant tube worms cluster around hydrothermal vents

Nematodes are ubiquitous pseudocoelomates which can parasite marine plants and animals.

Bloodworms are typically found on the bottom of shallow marine waters

Echinoderms

Echinoderms (Greek for *spiny skin*) is a phylum which contains only marine invertebrates. The adults are recognizable by their radial symmetry (usually five-point) and include starfish, sea urchins, sand dollars, and sea cucumbers, as well as the sea lilies. Echinoderms are found at every ocean depth, from the intertidal zone to the abyssal zone. The phylum contains about 7000 living species, making it the second-largest grouping of deuterostomes (a superphylum), after the chordates (which include the vertebrates, such as birds, fishes, mammals, and reptiles).

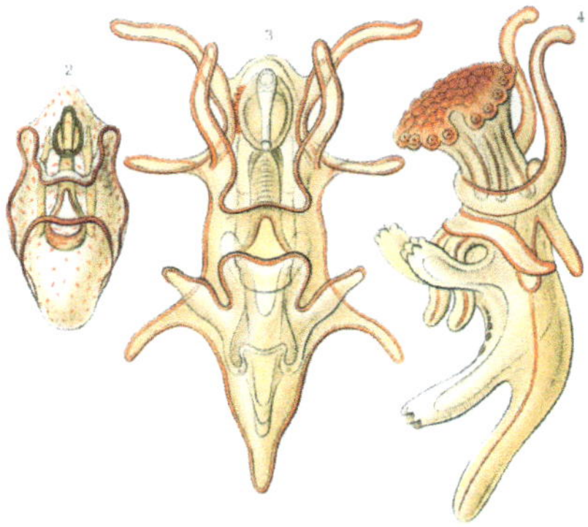

Starfish larvae are bilaterally symmetric, whereas the adults have fivefold symmetry

Echinoderms are unique among animals in having bilateral symmetry at the larval stage, but fivefold symmetry (pentamerism, a special type of radial symmetry) as adults.

The echinoderms are important both biologically and geologically. Biologically, there are few other groupings so abundant in the biotic desert of the deep sea, as well as shallower oceans. Most echinoderms are able to regenerate tissue, organs, limbs, and reproduce asexually; in some cases, they can undergo complete regeneration from a single limb. Geologically, the value of echinoderms is in their ossified skeletons, which are major contributors to many limestone formations, and can provide valuable clues as to the geological environment. They were the most used species in regenerative research in the 19th and 20th centuries. Further, it is held by some scientists that the radiation of echinoderms was responsible for the Mesozoic Marine Revolution.

		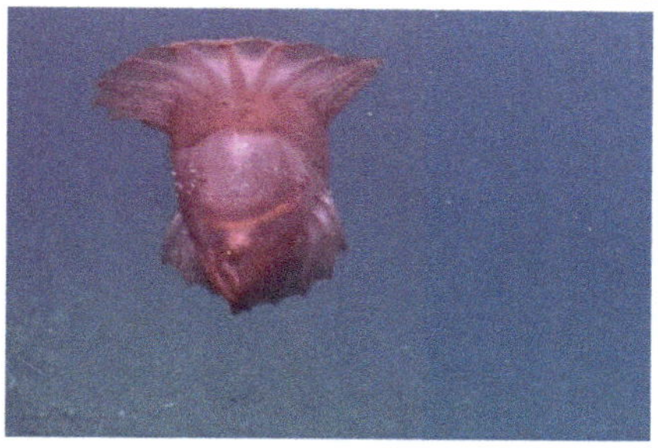
Sea cucumbers filter feed on plankton and suspended solids	Echinoderm literally means "spiny skin", as this water melon sea urchin illustrates	Benthopelagic sea cucumbers can lift off the seafloor and journey as much as 1,000 m (3,300 ft) up the water column

Aside from the hard-to-classify *Arkarua* (a Precambrian animal with echinoderm-like pentamerous radial symmetry), the first definitive members of the phylum appeared near the start of the Cambrian.

Marine Molluscs

Molluscs (Latin for *soft*) form a phylum with about 85,000 extant recognized species. By species count they are the largest marine phylum, comprising about 23% of all the named marine organisms. Molluscs have more varied forms than other invertebrate phylums. They are highly diverse, not just in size and in anatomical structure, but also in behaviour and in habitat. The majority of species still live in the oceans, from the seashores to the abyssal zone, but some form a significant part of the freshwater fauna and the terrestrial ecosystems.

Reconstruction of an ammonite, a highly successful early cephalopod that first appeared in the Devonian (about 400 mya). They became extinct during the same extinction event that killed the land dinosaurs (about 66 mya).

The mollusc phylum is divided into 9 or 10 taxonomic classes, two of which are extinct. These classes include gastropods, bivalves and cephalopods, as well as other lesser-known but distinctive classes. Gastropods with protective shells are referred to as snails, whereas gastropods without protective shells are referred to as slugs. Gastropods are by far the most numerous molluscs in terms of classified species, accounting for 80% of the total. Bivalves include clams, oysters, cockles, mussels, scallops, and numerous other families. There are about 8,000 marine bivalves species (including brackish water and estuarine species), and about 1,200 freshwater species. Cephalopod include octopus, squid and cuttlefish. They are found in all oceans, and neurologically are the most advanced of the invertebrates. About 800 living species of marine cephalopods have been identified, and an estimated 11,000 extinct taxa have been described. There are no fully freshwater cephalopods.

The nautilus is a living fossil little changed since it evolved 500 million years ago as one of the first cephalopods.	Marine gastropods are sea snails or sea slugs. This nudibranch is a sea slug.	Molluscs usually have eyes. Bordering the edge of the mantle of a scallop, a bivalve mollusc, can be over 100 simple eyes.

Molluscs have such diverse shapes that many textbooks base their descriptions of molluscan anatomy on a generalized or *hypothetical ancestral mollusc*. This generalized mollusc is unsegmented and bilaterally symmetrical with an underside consisting of a single muscular foot. Beyond that it has three further key features. Firstly, it has a muscular cloak called a mantle covering its viscera and containing a significant cavity used for breathing and excretion. A shell secreted by the mantle covers the upper surface. Secondly (apart from bivalves) it has a rasping tongue called a radula used for feeding. Thirdly, it has a nervous system including a complex digestive system using microscopic, muscle-powered hairs called cilia to exude mucus. The generalized mollusc has two paired nerve cords (three in bivalves). The brain, in species that have one, encircles the esophagus. Most molluscs have eyes and all have sensors detecting chemicals, vibrations, and touch. The simplest type of molluscan reproductive system relies on external fertilization, but more complex variations occur. All produce eggs, from which may emerge trochophore larvae, more complex veliger larvae, or miniature adults. The depiction is rather similar to modern monoplacophorans, and some suggest it may resemble very early molluscs.

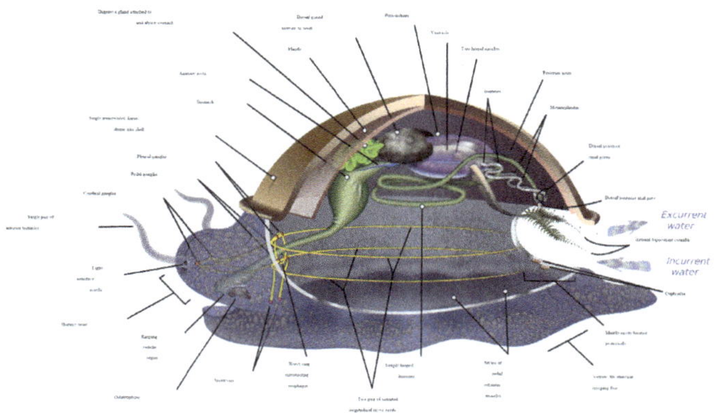

Generalized or *hypothetical ancestral mollusc*

Good evidence exists for the appearance of marine gastropods, cephalopods and bivalves in the Cambrian period 541 to 485.4 million years ago. However, the evolutionary history both of molluscs' emergence from the ancestral Lophotrochozoa and of their diversification into the well-known living and fossil forms are still subjects of vigorous debate among scientists.

Marine Arthropods

Arthropods (Greek for *jointed feet*) have an exoskeleton (external skeleton), a segmented body, and jointed appendages (paired appendages). They form a phylum which includes insects, arachnids, myriapods, and crustaceans. Arthropods are characterized by their jointed limbs and cuticle made of chitin, often mineralised with calcium carbonate. The arthropod body plan consists of segments, each with a pair of appendages. The rigid cuticle inhibits growth, so arthropods replace it periodically by moulting. Their versatility has enabled them to become the most species-rich members of all ecological guilds in most environments.

Marine arthropods range in size from the microscopic crustacean *Stygotantulus* to the Japanese spider crab. Arthropods' primary internal cavity is a hemocoel, which accommodates their internal organs, and through which their haemolymph - analogue of blood - circulates; they have open circulatory systems. Like their exteriors, the internal organs of arthropods are generally built of repeated segments. Their nervous system is "ladder-like", with paired ventral nerve cords running through all segments and forming paired ganglia in each segment. Their heads are formed by fusion of varying numbers of segments, and their brains are formed by fusion of the ganglia of these segments and encircle the esophagus. The respiratory and excretory systems of arthropods vary, depending as much on their environment as on the subphylum to which they belong.

Their vision relies on various combinations of compound eyes and pigment-pit ocelli: in most species the ocelli can only detect the direction from which light is coming, and the compound eyes are the main source of information, but the main eyes of spiders are ocelli that can form images and, in a few cases, can swivel to track prey. Arthropods also have a wide range of chemical and mechanical sensors, mostly based on modifications of the many setae (bristles) that project through their cuticles. Arthropods' methods of reproduction and development are diverse; all terrestrial species use internal fertilization, but this is often by indirect transfer of the sperm via an appendage or the ground, rather than by direct injection. Marine species all lay eggs and use either internal or external fertilization. Arthropod hatchlings vary from miniature adults to grubs that lack jointed limbs and eventually undergo a total metamorphosis to produce the adult form.

Crustaceans		
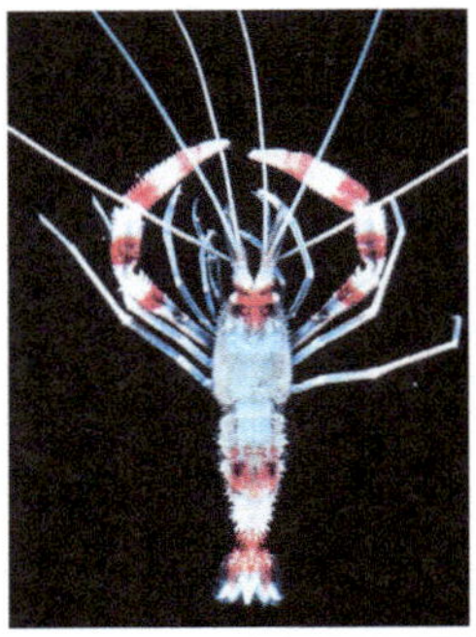	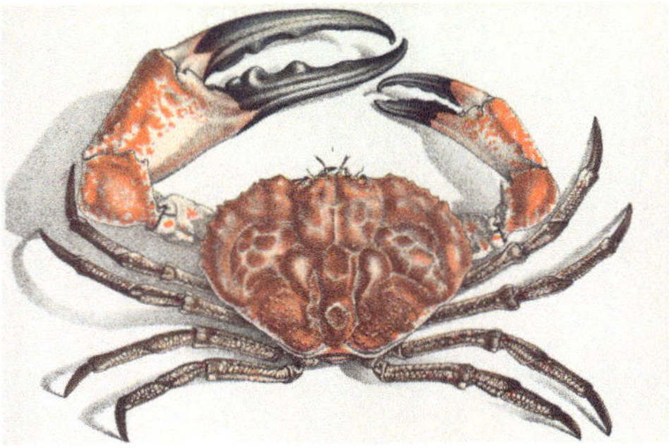	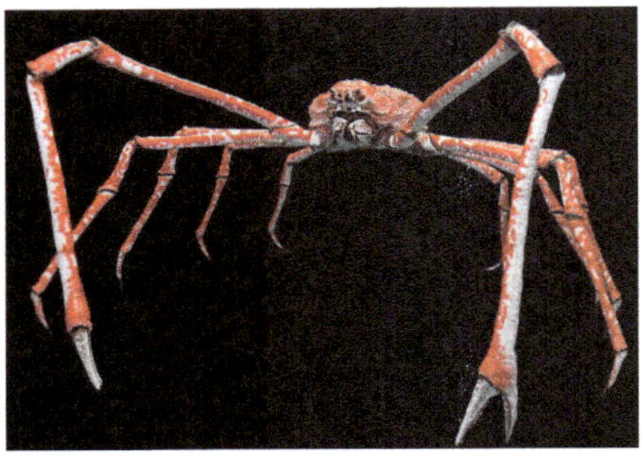
The banded cleaner shrimp is a crustacean common in the tropics.	The Tasmanian giant crab is long-lived and slow-growing, making it vulnerable to overfishing.	The Japanese spider crab has the longest leg span of any arthropod.

Many crustaceans are very small, like this tiny amphipod, and make up a significant part of the ocean's zooplankton

The evolutionary ancestry of arthropods dates back to the Cambrian period. The group is generally regarded as monophyletic, and many analyses support the placement of arthropods with cycloneuralians (or their constituent clades) in a superphylum Ecdysozoa. Overall however, the basal relationships of Metazoa are not yet well resolved. Likewise, the relationships between various arthropod groups are still actively debated.

Other Phyla

- Tardigrade, Lobopodia, (Onychophora)
- Non-craniate (non-vertebrate) chordates: Cephalochordate, Tunicata and *Haikouella*. These invertebrates are close relatives of the vertebrates.
- Non-craniate chordates are close relatives of vertebrates

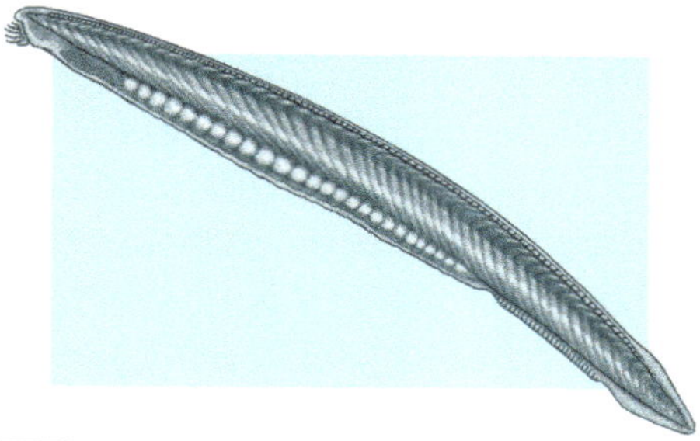	
The lancelet, a small translucent fish-like Cephalochordate, is the closest living invertebrate relative of the vertebrates.	Fluorescent-colored sea squirts, *Rhopalaea crassa*. Tunicates may provide clues to vertebrate (and therefore human) ancestry.
	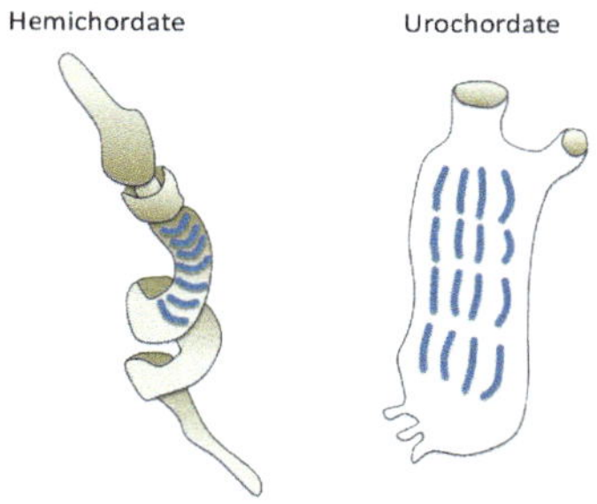
Salp chain	Gill slits in an acorn worm (left) and tunicate (right)

Minerals from Sea Water

There are a number of marine invertebrates that use minerals that are present in the sea in such minute quantities that they were undetectable until the advent of spectroscopy. Vanadium is concentrated by some tunicates for use in their blood cells to a level ten million times that of the surrounding seawater. Other tunicates similarly concentrate niobium and tantalum. Lobsters use copper in their respiratory pigment hemocyanin, despite the proportion of this metal in seawater being minute. Although these elements are present in vast quantities in the ocean, their extraction by man is not economic.

Cambrian Explosion

The Cambrian explosion or Cambrian radiation was the relatively short evolutionary event, beginning around 541 million years ago in the Cambrian period, during which most major animal phyla appeared, as indicated by the fossil record. Lasting for about the next 20–25 million years, it resulted in the divergence of most modern metazoan phyla. Additionally, the event was accompanied by major diversification of other organisms. Prior to the Cambrian explosion, most organisms were simple, composed of individual cells occasionally organized into colonies. Over the following 70 to 80 million years, the rate of diversification accelerated by an order of magnitude and the diversity of life began to resemble that of today. Almost all present animal phyla appeared during this period. There is strong evidence for species of Cnidaria and Porifera existing in the Ediacaran and possible members of Porifera even before that during the Cryogenian. Bryozoans don't appear in the fossil record until after the Cambrian, in the Lower Ordovician.

The Cambrian explosion has generated extensive scientific debate. The seemingly rapid appearance of fossils in the "Primordial Strata" was noted by William Buckland in the 1840s, and in 1859 Charles Darwin discussed it as one of the main objections that could be made against the theory of evolution by natural selection. The long-running puzzlement about the appearance of the Cambrian fauna, seemingly abruptly, without precursor, centers on three key points: whether there really was a mass diversification of complex organisms over a relatively short period of time during the early Cambrian; what might have caused such rapid change; and what it would imply about the origin of animal life. Interpretation is difficult due to a limited supply of evidence, based mainly on an incomplete fossil record and chemical signatures remaining in Cambrian rocks.

Phylogenetic analysis has been used to support the view that during the Cambrian explosion, metazoans (multi-celled animals) evolved monophyletically from a single common ancestor: flagellated colonial protists similar to modern choanoflagellates.

History and Significance

The first discovered Cambrian fossils were trilobites, described by Edward Lhuyd, the curator of Oxford Museum, in 1698. Although their evolutionary importance was not known, on the basis of their old age, William Buckland (1784–1856) realised that a dramatic step-change in the fossil record had occurred around the base of what we now call the Cambrian. Nineteenth-century geologists such as Adam Sedgwick and Roderick Murchison used the fossils for dating rock strata, specifically for establishing the Cambrian and Silurian periods. By 1859, leading geologists including Roderick Murchison, were convinced that what was then called the lowest Silurian stratum showed the origin of life on Earth, though others, including Charles Lyell, differed. In *On the Origin of Species*, Charles Darwin considered this sudden appearance of a solitary group of trilobites, with no apparent antecedents, and absence of other fossils, to be "undoubtedly of the gravest nature" among the difficulties in his theory of natural selection. He reasoned that earlier seas had swarmed with living creatures, but that their fossils had not been found due to the imperfections of the fossil record. In the sixth edition of his book, he stressed his problem further as:

To the question why we do not find rich fossiliferous deposits belonging to these assumed earliest periods prior to the Cambrian system, I can give no satisfactory answer.

American paleontologist Charles Walcott, who studied the Burgess Shale fauna, proposed that an interval of time, the "Lipalian", was not represented in the fossil record or did not preserve fossils, and that the ancestors of the Cambrian animals evolved during this time.

Earlier fossil evidence has since been found. The earliest claim is that the history of life on earth goes back 3,850 million years: Rocks of that age at Warrawoona, Australia, were claimed to contain fossil stromatolites, stubby pillars formed by colonies of microorganisms. Fossils (*Grypania*) of more complex eukaryotic cells, from which all animals, plants, and fungi are built, have been found in rocks from 1,400 million years ago, in China and Montana. Rocks dating from 580 to 543 million years ago contain fossils of the Ediacara biota, organisms so large that they are likely multicelled, but very unlike any modern organism. In 1948, Preston Cloud argued that a period of "eruptive" evolution occurred in the Early Cambrian, but as recently as the 1970s, no sign was seen of how the 'relatively' modern-looking organisms of the Middle and Late Cambrian arose.

Opabinia made the largest single contribution to modern interest in the Cambrian explosion.

The intense modern interest in this "Cambrian explosion" was sparked by the work of Harry B. Whittington and colleagues, who, in the 1970s, reanalysed many fossils from the Burgess Shale and concluded that several were complex, but different from any living animals. The most common organism, *Marrella*, was clearly an arthropod, but not a member of any known arthropod class. Organisms such as the five-eyed *Opabinia* and spiny slug-like *Wiwaxia* were so different from anything else known that Whittington's team assumed they must represent different phyla, seemingly unrelated to anything known today. Stephen Jay Gould's popular 1989 account of this work, *Wonderful Life*, brought the matter into the public eye and raised questions about what the explosion represented. While differing significantly in details, both Whittington and Gould proposed that all modern animal phyla had appeared almost simultaneously in a rather short span of geological period. This view led to the modernization of Darwin's tree of life and the theory of punctuated equilibrium, which Eldredge and Gould developed in the early 1970s and which views evolution as long intervals of near-stasis "punctuated" by short periods of rapid change.

Other analyses, some more recent and some dating back to the 1970s, argue that complex animals similar to modern types evolved well before the start of the Cambrian.

Dating the Cambrian

Radiometric dates for much of the Cambrian, obtained by analysis of radioactive elements contained within rocks, have only recently become available, and for only a few regions.

Relative dating (*A* was before *B*) is often assumed sufficient for studying processes of evolution, but this, too, has been difficult, because of the problems involved in matching up rocks of the same age across different continents.

Therefore, dates or descriptions of sequences of events should be regarded with some caution until better data become available.

Body Fossils

Fossils of organisms' bodies are usually the most informative type of evidence. Fossilization is a rare event, and most fossils are destroyed by erosion or metamorphism before they can be observed. Hence, the fossil record is very incomplete, increasingly so as earlier times are considered. Despite this, they are often adequate to illustrate the broader patterns of life's history. Also, biases exist in the fossil record: different environments are more favourable to the preservation of different types of organism or parts of organisms. Further, only the parts of organisms that were already mineralised are usually preserved, such as the shells of molluscs. Since most animal species are soft-bodied, they decay before they can become fossilised. As a result, although 30-plus phyla of living animals are known, two-thirds have never been found as fossils.

This *Marrella* specimen illustrates how clear and detailed the fossils from the Burgess Shale Lagerstätte are.

The Cambrian fossil record includes an unusually high number of lagerstätten, which preserve soft tissues. These allow paleontologists to examine the internal anatomy of animals, which in other sediments are only represented by shells, spines, claws, etc. – if they are preserved at all. The most significant Cambrian lagerstätten are the early Cambrian Maotianshan shale beds of Chengjiang (Yunnan, China) and Sirius Passet (Greenland); the middle Cambrian Burgess Shale (British Columbia, Canada); and the late Cambrian Orsten (Sweden) fossil beds.

While lagerstätten preserve far more than the conventional fossil record, they are far from complete. Because lagerstätten are restricted to a narrow range of environments (where soft-bodied organisms can be preserved very quickly, e.g. by mudslides), most animals are probably not represented; further, the exceptional conditions that create lagerstätten probably do not represent normal living conditions. In addition, the known Cambrian lagerstätten are rare and difficult to date, while Precambrian lagerstätten have yet to be studied in detail.

The sparseness of the fossil record means that organisms usually exist long before they are found in the fossil record – this is known as the Signor–Lipps effect.

Trace Fossils

Trace fossils consist mainly of tracks and burrows, but also include coprolites (fossil feces) and marks left by feeding. Trace fossils are particularly significant because they represent a data source that is not limited to animals with easily fossilized hard parts, and reflects organisms' behaviour.

Also, many traces date from significantly earlier than the body fossils of animals that are thought to have been capable of making them. While exact assignment of trace fossils to their makers is generally impossible, traces may, for example, provide the earliest physical evidence of the appearance of moderately complex animals (comparable to earthworms).

Rusophycus and other trace fossils from the Gog Group, Middle Cambrian, Lake Louise, Alberta, Canada

Geochemical Observations

Several chemical markers indicate a drastic change in the environment around the start of the Cambrian. The markers are consistent with a mass extinction, or with a massive warming resulting from the release of methane ice. Such changes may reflect a cause of the Cambrian explosion, although they may also have resulted from an increased level of biological activity – a possible result of the explosion. Despite these uncertainties, the geochemical evidence helps by making scientists focus on theories that are consistent with at least one of the likely environmental changes.

Phylogenetic Techniques

Cladistics is a technique for working out the "family tree" of a set of organisms. It works by the logic that, if groups B and C have more similarities to each other than either has to group A, then B and C are more closely related to each other than either is to A. Characteristics that are compared may be anatomical, such as the presence of a notochord, or molecular, by comparing sequences of DNA or protein. The result of a successful analysis is a hierarchy of clades – groups whose members are believed to share a common ancestor. The cladistic technique is sometimes problematic, as some features, such as wings or camera eyes, evolved more than once, convergently – this must be taken into account in analyses.

From the relationships, it may be possible to constrain the date that lineages first appeared. For instance, if fossils of B or C date to X million years ago and the calculated "family tree" says A was an ancestor of B and C, then A must have evolved more than X million years ago.

It is also possible to estimate how long ago two living clades diverged – i.e. about how long ago their last common ancestor must have lived – by assuming that DNA mutations accumulate at a constant rate. These "molecular clocks", however, are fallible, and provide only a very approximate timing: they are not sufficiently precise and reliable for estimating when the groups that feature in the Cambrian explosion first evolved, and estimates produced by different techniques vary by a factor of two. However, the clocks can give an indication of branching rate, and when combined with the constraints of the fossil record, recent clocks suggest a sustained period of diversification through the Ediacaran and Cambrian.

Explanation of Key Scientific Terms

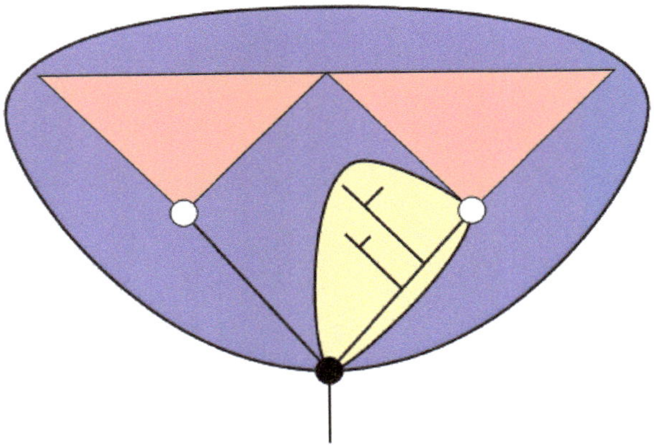

Stem groups

— = Lines of descent
= Basal node
= Crown node
= Total group
= Crown group
= Stem group

A phylum is the highest level in the Linnaean system for classifying organisms. Phyla can be thought of as groupings of animals based on general body plan. Despite the seemingly different external appearances of organisms, they are classified into phyla based on their internal and developmental organizations. For example, despite their obvious differences, spiders and barnacles both belong to the phylum Arthropoda, but earthworms and tapeworms, although similar in shape, belong to different phyla. As chemical and genetic testing becomes more accurate, previously hypothesised phyla are often entirely reworked.

A phylum is not a fundamental division of nature, such as the difference between electrons and protons. It is simply a very high-level grouping in a classification system created to describe all currently living organisms. This system is imperfect, even for modern animals: different books quote different numbers of phyla, mainly because they disagree about the classification of a huge number of worm-like species. As it is based on living organisms, it accommodates extinct organisms poorly, if at all.

The concept of stem groups was introduced to cover evolutionary "aunts" and "cousins" of living groups, and have been hypothesized based on this scientific theory. A crown group is a group of closely related living animals plus their last common ancestor plus all its descendants. A stem group is a set of offshoots from the lineage at a point earlier than the last common ancestor of the crown group; it is a relative concept, for example tardigrades are living animals that form a crown group in their own right, but Budd (1996) regarded them as also being a stem group relative to the arthropods.

The term *Triploblastic* means consisting of three layers, which are formed in the embryo, quite early in the animal's development from a single-celled egg to a larva or juvenile form. The innermost layer forms the digestive tract (gut); the outermost forms skin; and the middle one forms muscles and all the internal organs except the digestive system. Most types of living animal are triploblastic – the best-known exceptions are Porifera (sponges) and Cnidaria (jellyfish, sea anemones, etc.).

The bilaterians are animals that have right and left sides at some point in their life histories. This implies that they have top and bottom surfaces and, importantly, distinct front and back ends. All known bilaterian animals are triploblastic, and all known triploblastic animals are bilaterian. Living echinoderms (sea stars, sea urchins, sea cucumbers, etc.) 'look' radially symmetrical (like wheels) rather than bilaterian, but their larvae exhibit bilateral symmetry and some of the earliest echinoderms may have been bilaterally symmetrical. Porifera and Cnidaria are radially symmetrical, not bilaterian, and not triploblastic.

The term *Coelomate* means having a body cavity (coelom) containing the internal organs. Most of the phyla featured in the debate about the Cambrian explosion are coelomates: arthropods, annelid worms, molluscs, echinoderms, and chordates – the noncoelomate priapulids are an important exception. All known coelomate animals are triploblastic bilaterians, but some triploblastic bilaterian animals do not have a coelom – for example flatworms, whose organs are surrounded by unspecialized tissues.

Precambrian Life

Understanding of the Cambrian explosion relies upon knowing what was there beforehand – did the event herald the sudden appearance of a wide range of animals and behaviours, or did such things exist beforehand?

Evidence of Animals Around 1 Billion Years Ago

Stromatolites (Pika Formation, Middle Cambrian) near Helen Lake, Banff National Park, Canada

Changes in the abundance and diversity of some types of fossil have been interpreted as evidence for "attacks" by animals or other organisms. Stromatolites, stubby pillars built by colonies of microorganisms, are a major constituent of the fossil record from about 2,700 million years ago, but their abundance and diversity declined steeply after about 1,250 million years ago. This decline has been attributed to disruption by grazing and burrowing animals.

Modern stromatolites in Hamelin Pool Marine Nature Reserve, Western Australia

Our picture of Precambrian marine diversity is dominated by small fossils known as acritarchs. This term describes almost any small organic walled fossil – from the egg cases of small metazoans to resting cysts of many different kinds of green algae. After appearing around 2,000 million years ago, acritarchs underwent a boom around 1,000 million years ago, increasing in abundance, diversity, size, complexity of shape, and especially size and number of spines. Their increasingly spiny forms in the last 1 billion years may indicate an increased need for defence against predation. Other groups of small organisms from the Neoproterozoic era also show signs of antipredator defenses. A consideration of taxon longevity appears to support an increase in predation pressure around this time. In general, the fossil record shows a very slow appearance of these lifeforms in the Precambrian, with many cyanobacterial species making up much of the underlying sediment.

Fossils of the Doushantuo Formation

The layers of the Doushantuo formation from around 580 million year old harbour microscopic fossils that may represent early bilaterians. Some have been described as animal embryos and eggs, although some may represent the remains of giant bacteria. Another fossil, *Vernanimalcula*, has been interpreted as a coelomate bilaterian, but may simply be an infilled bubble.

These fossils form the earliest hard-and-fast evidence of animals, as opposed to other predators.

Burrows

The traces of organisms moving on and directly underneath the microbial mats that covered the Ediacaran sea floor are preserved from the Ediacaran period, about 565 million years ago. They were probably made by organisms resembling earthworms in shape, size, and how they moved. The burrow-makers have never been found preserved, but, because they would need a head and a tail, the burrowers probably had bilateral symmetry – which would in all probability make them bilaterian animals. They fed above the sediment surface, but were forced to burrow to avoid predators.

An Ediacaran trace fossil, made when an organism burrowed below a microbial mat.

Around the start of the Cambrian (about 542 million years ago), many new types of traces first appear, including well-known vertical burrows such as *Diplocraterion* and *Skolithos*, and traces normally attributed to arthropods, such as *Cruziana* and *Rusophycus*. The vertical burrows indicate that worm-like animals acquired new behaviours, and possibly new physical capabilities. Some Cambrian trace fossils indicate that their makers possessed hard exoskeletons, although they were not necessarily mineralised.

Burrows provide firm evidence of complex organisms; they are also much more readily preserved than body fossils, to the extent that the absence of trace fossils has been used to imply the genuine absence of large, motile, bottom-dwelling organisms. They provide a further line of evidence to show that the Cambrian explosion represents a real diversification, and is not a preservational artefact.

Indeed, as burrowing became established, it allowed an explosion of its own, for as burrowers disturbed the sea floor, they aerated it, mixing oxygen into the toxic muds. This made the bottom sediments more hospitable, and allowed a wider range of organisms to inhabit them – creating new niches and the scope for higher diversity.

Ediacaran Organisms

Dickinsonia costata, an Ediacaran organism of unknown affinity, with a quilted appearance

At the start of the Ediacaran period, much of the acritarch fauna, which had remained relatively unchanged for hundreds of millions of years, became extinct, to be replaced with a range of new, larger species, which would prove far more ephemeral. This radiation, the first in the fossil record, is followed soon after by an array of unfamiliar, large, fossils dubbed the Ediacara biota, which flourished for 40 million years until the start of the Cambrian. Most of this "Ediacara biota" were at least a few centimeters long, significantly larger than any earlier fossils. The organisms form three distinct assemblages, increasing in size and complexity as time progressed.

Many of these organisms were quite unlike anything that appeared before or since, resembling discs, mud-filled bags, or quilted mattresses – one palæontologist proposed that the strangest organisms should be classified as a separate kingdom, Vendozoa.

Fossil of *Kimberella*, a triploblastic bilaterian, and possibly a mollusc

At least some may have been early forms of the phyla at the heart of the "Cambrian explosion" debate, having been interpreted as early molluscs (*Kimberella*), echinoderms (*Arkarua*); and arthropods (*Spriggina*, *Parvancorina*). Still, debate exists about the classification of these specimens, mainly because the diagnostic features that allow taxonomists to classify more recent organisms, such as similarities to living organisms, are generally absent in the ediacarans. However, there seems little doubt that *Kimberella* was at least a triploblastic bilaterian animal. These organisms are central to the debate about how abrupt the Cambrian explosion was. If some were early members of the animal phyla seen today, the "explosion" looks a lot less sudden than if all these organisms represent an unrelated "experiment", and were replaced by the animal kingdom fairly soon thereafter (40M years is "soon" by evolutionary and geological standards).

Beck Spring Dolomite

Paul Knauth, a geologist at Arizona State University, maintains that photosynthesizing organisms such as algae, may have grown over a 750- to 800-million-year-old formation in Death Valley known as the Beck Spring Dolomite. In the early 1990s, samples from this 1,000-foot thick layer of dolomite revealed that the region housed flourishing mats of photosynthesizing, unicellular life forms which antedated the Cambrian explosion.

Microfossils have been unearthed from holes riddling the otherwise barren surface of the dolomite. These geochemical and microfossil findings support the idea that during the Precambrian period, complex life evolved both in the oceans and on land. Knauth contends that animals may well have had their origins in freshwater lakes and streams, and not in the oceans.

Some 30 years later, a number of studies have documented an abundance of geochemical and microfossil evidence showing that life covered the continents as far back as 2.2 billion years ago.

Many paleobiologists now accept the idea that simple life forms existed on land during the Precambrian, but are opposed to the more radical idea that multicellular life thrived on land more than 600 million years ago.

Ediacaran–Early Cambrian Skeletonisation

The first Ediacaran and lowest Cambrian (Nemakit-Daldynian) skeletal fossils represent tubes and problematic sponge spicules. The oldest sponge spicules are monaxon siliceous, aged around 580 million years ago, known from the Doushantou Formation in China and from deposits of the same age in Mongolia, although the interpretation of these fossils as spicules has been challenged. In the late Ediacaran-lowest Cambrian, numerous tube dwellings of enigmatic organisms appeared. It was organic-walled tubes (e.g. *Saarina*) and chitinous tubes of the sabelliditids (e.g. *Sokoloviina, Sabellidites, Paleolina*) that prospered up to the beginning of the Tommotian. The mineralized tubes of *Cloudina, Namacalathus, Sinotubulites*, and a dozen more of the other organisms from carbonate rocks formed near the end of the Ediacaran period from 549 to 542 million years ago, as well as the triradially symmetrical mineralized tubes of anabaritids (e.g. *Anabarites, Cambrotubulus*) from uppermost Ediacaran and lower Cambrian. Ediacaran mineralized tubes are often found in carbonates of the stromatolite reefs and thrombolites, i.e. they could live in an environment adverse to the majority of animals.

Although they are as hard to classify as most other Ediacaran organisms, they are important in two other ways. First, they are the earliest known calcifying organisms (organisms that built shells from calcium carbonate). Secondly, these tubes are a device to rise over a substrate and competitors for effective feeding and, to a lesser degree, they serve as armor for protection against predators and adverse conditions of environment. Some *Cloudina* fossils show small holes in shells. The holes possibly are evidence of boring by predators sufficiently advanced to penetrate shells. A possible "evolutionary arms race" between predators and prey is one of the hypotheses that attempt to explain the Cambrian explosion.

In the lowest Cambrian, the stromatolites were decimated. This allowed animals to begin colonization of warm-water pools with carbonate sedimentation. At first, it was anabaritids and *Protohertzina* (the fossilized grasping spines of chaetognaths) fossils. Such mineral skeletons as shells, sclerites, thorns, and plates appeared in uppermost Nemakit-Daldynian; they were the earliest species of halkierids, gastropods, hyoliths and other rare organisms. The beginning of the Tommotian has historically been understood to mark an explosive increase of the number and variety of fossils of molluscs, hyoliths, and sponges, along with a rich complex of skeletal elements of unknown animals, the first archaeocyathids, brachiopods, tommotiids, and others. This sudden increase is partially an artefact of missing strata at the Tommotian type section, and most of this fauna in fact began to diversify in a series of pulses through the Nemakit-Daldynian and into the Tommotian.

Some animals may already have had sclerites, thorns, and plates in the Ediacaran (e.g. *Kimberella* had hard sclerites, probably of carbonate), but thin carbonate skeletons cannot be fossilized in siliciclastic deposits. Older (~750 Ma) fossils indicate that mineralization long preceded the Cambrian, probably defending small photosynthetic algae from single-celled eukaryotic predators.

Cambrian Life

Trace Fossils

Trace fossils (burrows, etc.) are a reliable indicator of what life was around, and indicate a diversification of life around the start of the Cambrian, with the freshwater realm colonized by animals almost as quickly as the oceans.

Small Shelly Fauna

Fossils known as "small shelly fauna" have been found in many parts on the world, and date from just before the Cambrian to about 10 million years after the start of the Cambrian (the Nemakit-Daldynian and Tommotian ages). These are a very mixed collection of fossils: spines, sclerites (armor plates), tubes, archeocyathids (sponge-like animals), and small shells very like those of brachiopods and snail-like molluscs – but all tiny, mostly 1 to 2 mm long.

While small, these fossils are far more common than complete fossils of the organisms that produced them; crucially, they cover the window from the start of the Cambrian to the first lagerstätten: a period of time otherwise lacking in fossils. Hence, they supplement the conventional fossil record and allow the fossil ranges of many groups to be extended.

Early Cambrian Trilobites and Echinoderms

A fossilized trilobite, an ancient type of arthropod: This specimen, from the Burgess Shale, preserves "soft parts" – the antennae and legs.

The earliest trilobite fossils are about 530 million years old, but the class was already quite diverse and worldwide, suggesting they had been around for quite some time. The fossil record of trilobites began with the appearance of trilobites with mineral exoskeletons – not from the time of their origin.

The earliest generally accepted echinoderm fossils appeared a little bit later, in the Late Atdabanian; unlike modern echinoderms, these early Cambrian echinoderms were not all radially symmetrical.

These provide firm data points for the "end" of the explosion, or at least indications that the crown groups of modern phyla were represented.

Burgess Shale Type Faunas

The Burgess Shale and similar lagerstätten preserve the soft parts of organisms, which provide a wealth of data to aid in the classification of enigmatic fossils. It often preserved complete specimens of organisms only otherwise known from dispersed parts, such as loose scales or isolated mouthparts. Further, the majority of organisms and taxa in these horizons are entirely soft-bodied, hence absent from the rest of the fossil record. Since a large part of the ecosystem is preserved, the ecology of the community can also be tentatively reconstructed. However, the assemblages may represent a "museum": a deep-water ecosystem that is evolutionarily "behind" the rapidly diversifying fauna of shallower waters.

Because the lagerstätten provide a mode and quality of preservation that is virtually absent outside of the Cambrian, many organisms appear completely different from anything known from the conventional fossil record. This led early workers in the field to attempt to shoehorn the organisms into extant phyla; the shortcomings of this approach led later workers to erect a multitude of new phyla to accommodate all the oddballs. It has since been realised that most oddballs diverged from lineages before they established the phyla known today slightly different designs, which were fated to perish rather than flourish into phyla, as their cousin lineages did.

The preservational mode is rare in the preceding Ediacaran period, but those assemblages known show no trace of animal life perhaps implying a genuine absence of macroscopic metazoans.

Early Cambrian Crustaceans

Crustaceans, one of the four great modern groups of arthropods, are very rare throughout the Cambrian. Convincing crustaceans were once thought to be common in Burgess Shale-type biotas, but none of these individuals can be shown to fall into the crown group of "true crustaceans". The Cambrian record of crown-group crustaceans comes from microfossils. The Swedish Orsten horizons contain later Cambrian crustaceans, but only organisms smaller than 2 mm are preserved. This restricts the data set to juveniles and miniaturised adults.

A more informative data source is the organic microfossils of the Mount Cap formation, Mackenzie Mountains, Canada. This late Early Cambrian assemblage (510 to 515 million years ago) consists of microscopic fragments of arthropods' cuticle, which is left behind when the rock is dissolved with hydrofluoric acid. The diversity of this assemblage is similar to that of modern crustacean faunas. Most interestingly, analysis of fragments of feeding machinery found in the formation shows that it was adapted to feed in a very precise and refined fashion. This contrasts with most other early Cambrian arthropods, which fed messily by shovelling anything they could get their feeding appendages on into their mouths. This sophisticated and specialised feeding machinery belonged to a large (about 30 cm) organism, and would have provided great potential for diversification; specialised feeding apparatus allows a number of different approaches to feeding and development, and creates a number of different approaches to avoid being eaten.

Early Ordovician Radiation

After an extinction at the Cambrian-Ordovician boundary, another radiation occurred, which established the taxa that would dominate the Palaeozoic.

During this radiation, the total number of orders doubled, and families tripled, increasing marine diversity to levels typical of the Palaeozoic, and disparity to levels approximately equivalent to today's.

Stages

Different authors break the explosion down into stages in different ways.

Ed Landing recognizes three stages: Stage 1, spanning the Ediacaran-Cambrian boundary, corresponds to a diversification of biomineralizing animals and of deep and complex burrows; Stage 2 corresponds to the radiation of molluscs and stem-group Brachiopods (hyoliths and tommotiids), which apparently arose in intertidal waters; and Stage 3 sees the Atdabanian diversification of trilobites in deeper waters, but little change in the intertidal realm.

Graham Budd synthesises various schemes to produce a compatible view of the SSF record of the Cambrian explosion, divided slightly differently into four intervals: a "Tube world", lasting from 550 to 536 million years ago, spanning the E-€ boundary, dominated by Cloudina, Namacalathus ans pseudoconodont-type element; a "Sclerite world", seeing the rise of halkieriids, tommotiids, and hyoliths, lasting to the end of the Fortunian (c. 525 Ma); a brachiopod world, perhaps corresponding to the as yet unratified Cambrian Stage 2; and Trilobite World, kicking off in Stage 3.

Complementary to the shelly fossil record, trace fossils can be divided into five subdivisions: "Flat world" (late Ediacaran), with traces restricted to the sediment surface; Protreozoic III (after Jensen), with increasing complexity; *pedum* world, initiated at the base of the Cambrian with the base of the *T.pedum* zone; *Rusophycus* world, spanning 536 to 521 million years ago and thus corresponding exactly to the periods of Sclerite World and Brachiopod World under the SSF paradigm; and *Cruziana* world, with an obvious correspondence to Trilobite World.

How Real was the Explosion?

The fossil record as Darwin knew it seemed to suggest that the major metazoan groups appeared in a few million years of the early to mid-Cambrian, and even in the 1980s, this still appeared to be the case.

However, evidence of Precambrian Metazoa is gradually accumulating. If the Ediacaran *Kimberella* was a mollusc-like protostome (one of the two main groups of coelomates), the protostome and deuterostome lineages must have split significantly before 550 million years ago (deuterostomes are the other main group of coelomates). Even if it is not a protostome, it is widely accepted as a bilaterian. Since fossils of rather modern-looking cnidarians (jellyfish-like organisms) have been found in the Doushantuo lagerstätte, the cnidarian and bilaterian lineages must have diverged well over 580 million years ago.

Trace fossils and predatory borings in *Cloudina* shells provide further evidence of Ediacaran animals. Some fossils from the Doushantuo formation have been interpreted as embryos and one (*Vernanimalcula*) as a bilaterian coelomate, although these interpretations are not universally accepted. Earlier still, predatory pressure has acted on stromatolites and acritarchs since around 1,250 million years ago.

The presence of Precambrian animals somewhat dampens the "bang" of the explosion; not only was the appearance of animals gradual, but their evolutionary radiation ("diversification") may also not have been as rapid as once thought. Indeed, statistical analysis shows that the Cambrian explosion was no faster than any of the other radiations in animals' history. However, it does seem that some innovations linked to the explosion – such as resistant armour – only evolved once in the animal lineage; this makes a lengthy Precambrian animal lineage harder to defend. Further, the conventional view that all the phyla arose in the Cambrian is flawed; while the phyla may have diversified in this time period, representatives of the crown groups of many phyla do not appear until much later in the Phanerozoic. Further, the mineralised phyla that form the basis of the fossil record may not be representative of other phyla, since most mineralised phyla originated in a benthic setting. The fossil record is consistent with a Cambrian explosion that was limited to the benthos, with pelagic phyla evolving much later.

Ecological complexity among marine animals increased in the Cambrian, as well later in the Ordovician. However, recent research has overthrown the once-popular idea that disparity was exceptionally high throughout the Cambrian, before subsequently decreasing. In fact, disparity remains relatively low throughout the Cambrian, with modern levels of disparity only attained after the early Ordovician radiation.

The diversity of many Cambrian assemblages is similar to today's, and at a high (class/phylum) level, diversity is thought by some to have risen relatively smoothly through the Cambrian, stabilizing somewhat in the Ordovician. This interpretation, however, glosses over the astonishing and fundamental pattern of basal polytomy and phylogenetic telescoping at or near the Cambrian boundary, as seen in most major animal lineages. Thus Harry Blackmore Whittington's questions regarding the abrupt nature of the Cambrian explosion remain, and have yet to be satisfactorily answered.

Possible Causes

Despite the evidence that moderately complex animals (triploblastic bilaterians) existed before and possibly long before the start of the Cambrian, it seems that the pace of evolution was exceptionally fast in the early Cambrian. Possible explanations for this fall into three broad categories: environmental, developmental, and ecological changes. Any explanation must explain both the timing and magnitude of the explosion.

Changes in the Environment

Increase in Oxygen Levels

Earth's earliest atmosphere contained no free oxygen (O_2); the oxygen that animals breathe today, both in the air and dissolved in water, is the product of billions of years of photosynthesis. Cyanobacteria were the first organisms to evolve the ability to photosynthesize, introducing a steady supply of oxygen into the environment. Initially, oxygen levels did not increase substantially in the atmosphere. The oxygen quickly reacted with iron and other minerals in the surrounding rock and ocean water. Once a saturation point was reached for the reactions in rock and water, oxygen was able to exist as a gas in its diatomic form. Oxygen levels in the atmosphere increased substantially afterward. As a general trend, the concentration of oxygen in the atmosphere has risen gradually over about the last 2.5 billion years.

Oxygen levels seem to have a positive correlation with diversity in eukaryotes well before the Cambrian period. The last common ancestor of all extant eukaryotes is thought to have lived around 1.8 billion years ago. Around 800 million years ago, there was a notable increase in the complexity and number of eukaryotes species in the fossil record. Before the spike in diversity, eukaryotes are thought to have lived in highly sulfuric environments. Sulfide interferes with mitochondrial function in aerobic organisms, limiting the amount of oxygen that could be used to drive metabolism. Oceanic sulfide levels decreased around 800 million years ago, which supports the importance of oxygen in eukaryotic diversity.

The shortage of oxygen might well have prevented the rise of large, complex animals. The amount of oxygen an animal can absorb is largely determined by the area of its oxygen-absorbing surfaces (lungs and gills in the most complex animals; the skin in less complex ones); but, the amount needed is determined by its volume, which grows faster than the oxygen-absorbing area if an animal's size increases equally in all directions. An increase in the concentration of oxygen in air or water would increase the size to which an organism could grow without its tissues becoming starved of oxygen. However, members of the Ediacara biota reached metres in length tens of millions of years before the Cambrian explosion. Other metabolic functions may have been inhibited by lack of oxygen, for example the construction of tissue such as collagen, required for the construction of complex structures, or to form molecules for the construction of a hard exoskeleton. However, animals are not affected when similar oceanographic conditions occur in the Phanerozoic; there is no convincing correlation between oxygen levels and evolution, so oxygen may have been no more a prerequisite to complex life than liquid water or primary productivity.

Ozone Formation

The amount of ozone (O_3) required to shield Earth from biologically lethal UV radiation, wavelengths from 200 to 300 nanometers (nm), is believed to have been in existence around the Cambrian explosion. The presence of the ozone layer may have enabled the development of complex life and life on land, as opposed to life being restricted in the water.

Snowball Earth

In the late Neoproterozoic (extending into the early Ediacaran period), the Earth suffered massive glaciations in which most of its surface was covered by ice. This may have caused a mass extinction, creating a genetic bottleneck; the resulting diversification may have given rise to the Ediacara biota, which appears soon after the last "Snowball Earth" episode. However, the snowball episodes occurred a long time before the start of the Cambrian, and it is hard to see how so much diversity could have been caused by even a series of bottlenecks; the cold periods may even have *delayed* the evolution of large size organisms.

Increase in the Calcium Concentration of the Cambrian Seawater

Newer research suggests that volcanically active midocean ridges caused a massive and sudden surge of the calcium concentration in the oceans, making it possible for marine organisms to build skeletons and hard body parts. Alternatively a high influx of ions could have been provided by the widespread erosion that produced Powell's Great Unconformity.

Developmental Explanations

A range of theories are based on the concept that minor modifications to animals' development as they grow from embryo to adult may have been able to cause very large changes in the final adult form. The Hox genes, for example, control which organs individual regions of an embryo will develop into. For instance, if a certain *Hox* gene is expressed, a region will develop into a limb; if a different Hox gene is expressed in that region (a minor change), it could develop into an eye instead (a phenotypically major change).

Such a system allows a large range of disparity to appear from a limited set of genes, but such theories linking this with the explosion struggle to explain why the origin of such a development system should by itself lead to increased diversity or disparity. Evidence of Precambrian metazoans combines with molecular data to show that much of the genetic architecture that could feasibly have played a role in the explosion was already well established by the Cambrian.

This apparent paradox is addressed in a theory that focuses on the physics of development. It is proposed that the emergence of simple multicellular forms provided a changed context and spatial scale in which novel physical processes and effects were mobilized by the products of genes that had previously evolved to serve unicellular functions. Morphological complexity (layers, segments, lumens, appendages) arose, in this view, by self-organization.

Horizontal gene transfer has also been identified as a possible factor in the rapid acquisition of the biochemical capability of biomineralization among organisms during this period, based on evidence that the gene for a critical protein in the process was originally transferred from a bacterium into sponges.

Ecological Explanations

These focus on the interactions between different types of organism. Some of these hypotheses deal with changes in the food chain; some suggest arms races between predators and prey, and others focus on the more general mechanisms of coevolution. Such theories are well suited to explaining why there was a rapid increase in both disparity and diversity, but they must explain why the "explosion" happened when it did.

End-Ediacaran Mass Extinction

Evidence for such an extinction includes the disappearance from the fossil record of the Ediacara biota and shelly fossils such as *Cloudina*, and the accompanying perturbation in the $\delta^{13}C$ record.

Mass extinctions are often followed by adaptive radiations as existing clades expand to occupy the ecospace emptied by the extinction. However, once the dust had settled, overall disparity and diversity returned to the pre-extinction level in each of the Phanerozoic extinctions.

Evolution of Eyes

Andrew Parker has proposed that predator-prey relationships changed dramatically after eyesight evolved. Prior to that time, hunting and evading were both close-range affairs – smell, vibration, and touch were the only senses used. When predators could see their prey from a distance, new defensive strategies were needed. Armor, spines, and similar defenses may also have evolved in response to vision. He further observed that, where animals lose vision in unlighted environments such as caves, diversity of animal forms tends to decrease. Nevertheless, many scientists doubt that vision could have caused the explosion. Eyes may well have evolved long before the start of the Cambrian. It is also difficult to understand why the evolution of eyesight would have caused an explosion, since other senses, such as smell and pressure detection, can detect things at a greater distance in the sea than sight can; but the appearance of these other senses apparently did not cause an evolutionary explosion.

Arms Races between Predators and Prey

The ability to avoid or recover from predation often makes the difference between life and death, and is therefore one of the strongest components of natural selection. The pressure to adapt is stronger on the prey than on the predator: if the predator fails to win a contest, it loses a meal; if the prey is the loser, it loses its life.

But, there is evidence that predation was rife long before the start of the Cambrian, for example in the increasingly spiny forms of acritarchs, the holes drilled in *Cloudina* shells, and traces of burrowing to avoid predators. Hence, it is unlikely that the *appearance* of predation was the trigger for the Cambrian "explosion", although it may well have exhibited a strong influence on the body forms that the "explosion" produced. However, the intensity of predation does appear to have increased dramatically during the Cambrian as new predatory "tactics" (such as shell-crushing) emerged. This rise of predation during the Cambrian was confirmed by the temporal pattern of the median predator ratio at the scale of genus, in fossil communities covering the Cambrian and Ordovician periods, but this pattern is not correlated to diversification rate. This lack of correlation between predator ratio and diversification over the Cambrian and Ordovician suggests that predators did not trigger the large evolutionary radiation of animals during this interval. Thus the role of predators as triggerers of diversification may have been limited to the very beginning of the "Cambrian explosion".

Increase in Size and Diversity of Planktonic Animals

Geochemical evidence strongly indicates that the total mass of plankton has been similar to modern levels since early in the Proterozoic. Before the start of the Cambrian, their corpses and droppings were too small to fall quickly towards the seabed, since their drag was about the same as their weight. This meant they were destroyed by scavengers or by chemical processes before they reached the sea floor.

Mesozooplankton are plankton of a larger size. Early Cambrian specimens filtered microscopic plankton from the seawater. These larger organisms would have produced droppings and corpses that were large enough to fall fairly quickly. This provided a new supply of energy and nutrients to the mid-levels and bottoms of the seas, which opened up a huge range of new possible ways of life. If any of these remains sank uneaten to the sea floor they could be buried; this would have taken some carbon out of circulation, resulting in an increase in the concentration of breathable oxygen in the seas (carbon readily combines with oxygen).

The initial herbivorous mesozooplankton were probably larvae of benthic (seafloor) animals. A larval stage was probably an evolutionary innovation driven by the increasing level of predation at the seafloor during the Ediacaran period.

Metazoans have an amazing ability to increase diversity through coevolution. This means that an organism's traits can lead to traits evolving in other organisms; a number of responses are possible, and a different species can potentially emerge from each one. As a simple example, the evolution of predation may have caused one organism to develop a defence, while another developed motion to flee. This would cause the predator lineage to split into two species: one that was good at chasing prey, and another that was good at breaking through defences. Actual coevolution is somewhat more subtle, but, in this fashion, great diversity can arise: three quarters of living species are animals, and most of the rest have formed by coevolution with animals.

Ecosystem Engineering

Evolving organisms inevitably change the environment they evolve in. The Devonian colonization of land had planet-wide consequences for sediment cycling and ocean nutrients, and was likely linked to the Devonian mass extinction. A similar process may have occurred on smaller scales in the oceans, with, for example, the sponges filtering particles from the water and depositing them in the mud in a more digestible form; or burrowing organisms making previously unavailable resources available for other organisms.

Complexity Threshold

The explosion may not have been a significant evolutionary event. It may represent a threshold being crossed: for example a threshold in genetic complexity that allowed a vast range of morphological forms to be employed. This genetic threshold may have a correlation to the amount of oxygen available to organisms. Using oxygen for metabolism produces much more energy than anaerobic processes. Organisms that use more oxygen have the opportunity to produce more complex proteins, providing a template for further evolution. These proteins translate into larger, more complex structures that allow organisms better to adapt to their environments. With the help of oxygen, genes that code for these proteins could contribute to the expression of complex traits more efficiently. Access to a wider range of structures and functions would allow organisms to evolve in different directions, increasing the number of niches that could be inhabited. Furthermore, organisms had the opportunity to become more specialized in their own niches.

Uniqueness of the Explosion

The "Cambrian explosion" can be viewed as two waves of metazoan expansion into empty niches: first, a coevolutionary rise in diversity as animals explored niches on the Ediacaran sea floor, followed by a second expansion in the early Cambrian as they became established in the water column. The rate of diversification seen in the Cambrian phase of the explosion is unparalleled among marine animals: it affected all metazoan clades of which Cambrian fossils have been found. Later radiations, such as those of fish in the Silurian and Devonian periods, involved fewer taxa, mainly with very similar body plans. Although the recovery from the Permian-Triassic extinction started with about as few animal species as the Cambrian explosion, the recovery produced far fewer significantly new types of animals.

Whatever triggered the early Cambrian diversification opened up an exceptionally wide range of previously unavailable ecological niches. When these were all occupied, limited space existed for such wide-ranging diversifications to occur again, because strong competition existed in all niches and incumbents usually had the advantage. If a wide range of empty niches had continued, clades would be able to continue diversifying and become disparate enough for us to recognise them as different phyla; when niches are filled, lineages will continue to resemble one another long after they diverge, as limited opportunity exists for them to change their life-styles and forms.

There were two similar explosions in the evolution of land plants: after a cryptic history beginning about 450 million years ago, land plants underwent a uniquely rapid adaptive radiation during the Devonian period, about 400 million years ago. Furthermore, Angiosperms (flowering plants) originated and rapidly diversified during the Cretaceous period.

Annelid

The annelids (*Annelida*, from Latin *anellus*, "little ring"), also known as the ringed worms or segmented worms, are a large phylum, with over 17,000 extant species including ragworms, earthworms, and leeches. The species exist in and have adapted to various ecologies - some in marine environments as distinct as tidal zones and hydrothermal vents, others in fresh water, and yet others in moist terrestrial environments.

The annelids are bilaterally symmetrical, triploblastic, coelomate, invertebrate organisms. They also have parapodia for locomotion. Most textbooks still use the traditional division into polychaetes (almost all marine), oligochaetes (which include earthworms) and leech-like species. Cladistic research since 1997 has radically changed this scheme, viewing leeches as a sub-group of oligochaetes and oligochaetes as a sub-group of polychaetes. In addition, the Pogonophora, Echiura and Sipuncula, previously regarded as separate phyla, are now regarded as sub-groups of polychaetes. Annelids are considered members of the Lophotrochozoa, a "super-phylum" of protostomes that also includes molluscs, brachiopods, flatworms and nemerteans.

The basic annelid form consists of multiple segments. Each segment has the same sets of organs and, in most polychates, has a pair of parapodia that many species use for locomotion. Septa separate the segments of many species, but are poorly defined or absent in others, and Echiura and Sipuncula show no obvious signs of segmentation. In species with well-developed septa, the blood circulates entirely within blood vessels, and the vessels in segments near the front ends of these species are often built up with muscles that act as hearts. The septa of such species also enable them to change the shapes of individual segments, which facilitates movement by peristalsis ("ripples" that pass along the body) or by undulations that improve the effectiveness of the parapodia. In species with incomplete septa or none, the blood circulates through the main body cavity without any kind of pump, and there is a wide range of locomotory techniques – some burrowing species turn their pharynges inside out to drag themselves through the sediment.

Although many species can reproduce asexually and use similar mechanisms to regenerate after severe injuries, sexual reproduction is the normal method in species whose reproduction has been studied. The minority of living polychaetes whose reproduction and lifecycles are known produce trochophore larvae, that live as plankton and then sink and metamorphose into miniature adults. Oligochaetes are full hermaphrodites and produce a ring-like cocoon around their bodies, in which the eggs and hatchlings are nourished until they are ready to emerge.

Earthworms are Oligochaetes that support terrestrial food chains both as prey and in some regions are important in aeration and enriching of soil. The burrowing of marine polychaetes, which may constitute up to a third of all species in near-shore environments, encourages the development of ecosystems by enabling water and oxygen to penetrate the sea floor. In addition to improving soil fertility, annelids serve humans as food and as bait. Scientists observe annelids to monitor the quality of marine and fresh water. Although blood-letting is no longer in favor with doctors, some leech species are regarded as endangered species because they have been over-harvested for this purpose in the last few centuries. Ragworms' jaws are now being studied by engineers as they offer an exceptional combination of lightness and strength.

Since annelids are soft-bodied, their fossils are rare – mostly jaws and the mineralized tubes that some of the species secreted. Although some late Ediacaran fossils may represent annelids, the oldest known fossil that is identified with confidence comes from about 518 million years ago in the early Cambrian period. Fossils of most modern mobile polychaete groups appeared by the end of the Carboniferous, about 299 million years ago. Palaeontologists disagree about whether some body fossils from the mid Ordovician, about 472 to 461 million years ago, are the remains of oligochaetes, and the earliest indisputable fossils of the group appear in the Tertiary period, which began 65 million years ago.

Classification and Diversity

There are over 22,000 living annelid species, ranging in size from microscopic to the Australian giant Gippsland earthworm and *Amynthas mekongianus* (Cognetti, 1922), which can both grow up to 3 metres (9.8 ft) long. Although research since 1997 has radically changed scientists' views about the evolutionary family tree of the annelids, most textbooks use the traditional classification into the following sub-groups:

- Polychaetes (about 12,000 species). As their name suggests, they have multiple chetae ("hairs") per segment. Polychaetes have parapodia that function as limbs, and nuchal organs that are thought to be chemosensors. Most are marine animals, although a few species live in fresh water and even fewer on land.

An earthworm's clitellum

- Clitellates (about 10,000 species). These have few or no chetae per segment, and no nuchal organs or parapodia. However, they have a unique reproductive organ, the ring-shaped clitellum ("pack saddle") around their bodies, which produces a cocoon that stores and nourishes fertilized eggs until they hatch or, in moniligastrids, yolky eggs that provide nutrition for the embyros. The clitellates are sub-divided into:

- Oligochaetes ("with few hairs"), which includes earthworms. Oligochaetes have a sticky pad in the roof of the mouth. Most are burrowers that feed on wholly or partly decomposed organic materials.
- Hirudinea, whose name means "leech-shaped" and whose best known members are leeches. Marine species are mostly blood-sucking parasites, mainly on fish, while most freshwater species are predators. They have suckers at both ends of their bodies, and use these to move rather like inchworms.

The Archiannelida, minute annelids that live in the spaces between grains of marine sediment, were treated as a separate class because of their simple body structure, but are now regarded as polychaetes. Some other groups of animals have been classified in various ways, but are now widely regarded as annelids:

- Pogonophora / Siboglinidae were first discovered in 1914, and their lack of a recognizable gut made it difficult to classify them. They have been classified as a separate phylum, Pogonophora, or as two phyla, Pogonophora and Vestimentifera. More recently they have been re-classified as a family, Siboglinidae, within the polychaetes.
- The Echiura have a checkered taxonomic history: in the 19th century they were assigned to the phylum "Gephyrea", which is now empty as its members have been assigned to other phyla; the Echiura were next regarded as annelids until the 1940s, when they were classified as a phylum in their own right; but a molecular phylogenetics analysis in 1997 concluded that echiurans are annelids.
- Myzostomida live on crinoids and other echinoderms, mainly as parasites. In the past they have been regarded as close relatives of the trematode flatworms or of the tardigrades, but in 1998 it was suggested that they are a sub-group of polychaetes. However, another analysis in 2002 suggested that myzostomids are more closely related to flatworms or to rotifers and acanthocephales.

Distinguishing Features

No single feature distinguishes Annelids from other invertebrate phyla, but they have a distinctive combination of features. Their bodies are long, with segments that are divided externally by shallow ring-like constrictions called annuli and internally by septa ("partitions") at the same points, although in some species the septa are incomplete and in a few cases missing. Most of the segments contain the same sets of organs, although sharing a common gut, circulatory system and nervous system makes them inter-dependent. Their bodies are covered by a cuticle (outer covering) that does not contain cells but is secreted by cells in the skin underneath, is made of tough but flexible collagen and does not molt – on the other hand arthropods' cuticles are made of the more rigid α-chitin, and molt until the arthropods reach their full size. Most annelids have closed circulatory systems, where the blood makes its entire circuit via blood vessels.

Description

Segmentation

Most of an annelid's body consists of segments that are practically identical, having the same sets of internal organs and external chaetae (Greek, meaning "hair") and, in some species, appendages. However, the frontmost and rearmost sections are not regarded as true segments as they do not contain the standard sets of organs and do not develop in the same way as the true segments. The frontmost section, called the prostomium (Greek, meaning "in front of" and στομα meaning "mouth") contains the brain and sense organs, while the rearmost, called the pygidium (Greek, meaning "little tail") or periproct contains the anus, generally on the underside. The first section behind the prostomium, called the peristomium (Greek, meaning "around" and meaning "mouth"), is regarded by some zoologists as not a true segment, but in some polychaetes the peristomium has chetae and appendages like those of other segments.

The segments develop one at a time from a growth zone just ahead of the pygidium, so that an annelid's youngest segment is just in front of the growth zone while the peristomium is the oldest. This pattern is called teloblastic growth. Some groups of annelids, including all leeches, have fixed maximum numbers of segments, while others add segments throughout their lives.

The phylum's name is derived from the Latin word *annelus*, meaning "little ring".

Body Wall, Chetae and Parapodia

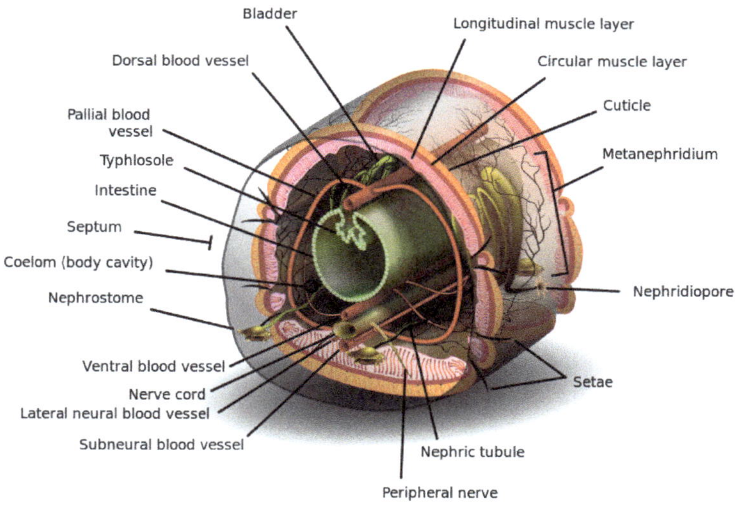

Annelids' cuticles are made of collagen fibers, usually in layers that spiral in alternating directions so that the fibers cross each other. These are secreted by the one-cell deep epidermis (outermost skin layer). A few marine annelids that live in tubes lack cuticles, but their tubes have a similar structure, and mucus-secreting glands in the epidermis protect their skins. Under the epidermis is the dermis, which is made of connective tissue, in other words a combination of cells and non-cellular materials such as collagen. Below this are two layers of muscles, which develop from the lining of the coelom (body cavity): circular muscles make a segment longer and slimmer when they contract, while under them are longitudinal muscles, usually four distinct strips, whose contractions make the segment shorter and fatter. Some annelids also have oblique internal muscles that connect the underside of the body to each side.

The setae ("hairs") of annelids project out from the epidermis to provide traction and other capabilities. The simplest are unjointed and form paired bundles near the top and bottom of each side of each segment. The parapodia ("limbs") of annelids that have them often bear more complex chetae at their tips – for example jointed, comb-like or hooked. Chetae are made of moderately flexible β-chitin and are formed by follicles, each of which has a chetoblast ("hair-forming") cell at the bottom and muscles that can extend or retract the cheta. The chetoblasts produce chetae by forming microvilli, fine hair-like extensions that increase the area available for secreting the cheta. When the cheta is complete, the microvilli withdraw into the chetoblast, leaving parallel tunnels that run almost the full length of the cheta. Hence annelids' chetae are structurally different from the setae ("bristles") of arthropods, which are made of the more rigid α-chitin, have a single internal cavity, and are mounted on flexible joints in shallow pits in the cuticle.

Nearly all polychaetes have parapodia that function as limbs, while other major annelid groups lack them. Parapodia are unjointed paired extensions of the body wall, and their muscles are derived from the circular muscles of the body. They are often supported internally by one or more large, thick chetae. The parapodia of burrowing and tube-dwelling polychaetes are often just ridges whose tips bear hooked chetae. In active crawlers and swimmers the parapodia are often divided into large upper and lower paddles on a very short trunk, and the paddles are generally fringed with chetae and sometimes with cirri (fused bundles of cilia) and gills.

Nervous System and Senses

The brain generally forms a ring round the pharynx (throat), consisting of a pair of ganglia (local control centers) above and in front of the pharynx, linked by nerve cords either side of the pharynx to another pair of ganglia just below and behind it. The brains of polychaetes are generally in the prostomium, while those of clitellates are in the peristomium or sometimes the first segment behind the peristomium. In some very mobile and active polychaetes the brain is enlarged and more complex, with visible hindbrain, midbrain and forebrain sections. The rest of the central nervous system is generally "ladder-like", consisting of a pair of nerve cords that run through the bottom part of the body and have in each segment paired ganglia linked by a transverse connection. From each segmental ganglion a branching system of local nerves runs into the body wall and then encircles the body. However, in most polychaetes the two main nerve cords are fused, and in the tube-dwelling genus *Owenia* the single nerve chord has no ganglia and is located in the epidermis.

As in arthropods, each muscle fiber (cell) is controlled by more than one neuron, and the speed and power of the fiber's contractions depends on the combined effects of all its neurons. Vertebrates have a different system, in which one neuron controls a group of muscle fibers. Most annelids' longitudinal nerve trunks include giant axons (the output signal lines of nerve cells). Their large diameter decreases their resistance, which allows them to transmit signals exceptionally fast. This enables these worms to withdraw rapidly from danger by shortening their bodies. Experiments have shown that cutting the giant axons prevents this escape response but does not affect normal movement.

The sensors are primarily single cells that detect light, chemicals, pressure waves and contact, and are present on the head, appendages (if any) and other parts of the body. Nuchal ("on the neck") organs are paired, ciliated structures found only in polychaetes, and are thought to be chemosensors. Some polychaetes also have various combinations of ocelli ("little eyes") that detect the direction from which light is coming and camera eyes or compound eyes that can probably form images. The compound eyes probably evolved independently of arthropods' eyes. Some tube-worms use ocelli widely spread over their bodies to detect the shadows of fish, so that they can quickly withdraw into their tubes. Some burrowing and tube-dwelling polychaetes have statocysts (tilt and balance sensors) that tell them which way is down. A few polychaete genera have on the undersides of their heads palps that are used both in feeding and as "feelers", and some of these also have antennae that are structurally similar but probably are used mainly as "feelers".

Coelom, Locomotion and Circulatory System

Most annelids have a pair of coelomata (body cavities) in each segment, separated from other segments by septa and from each other by vertical mesenteries. Each septum forms a sandwich with connective tissue in the middle and mesothelium (membrane that serves as a lining) from the preceding and following segments on either side. Each mesentery is similar except that the mesothelium is the lining of each of the pair of coelomata, and the blood vessels and, in polychaetes, the main nerve cords are embedded in it. The mesothelium is made of modified epitheliomuscular cells; in other words, their bodies form part of the epithelium but their bases extend to form muscle fibers in the body wall. The mesothelium may also form radial and circular muscles on the septa, and circular muscles around the blood vessels and gut. Parts of the mesothelium, especially on the outside of the gut, may also form chloragogen cells that perform similar functions to the livers of vertebrates: producing and storing glycogen and fat; producing the oxygen-carrier hemoglobin; breaking down proteins; and turning nitrogenous waste products into ammonia and urea to be excreted.

Many annelids move by peristalsis (waves of contraction and expansion that sweep along the body), or flex the body while using parapodia to crawl or swim. In these animals the septa enable the circular and longitudinal muscles to change the shape of individual segments, by making each segment a separate fluid-filled "balloon". However, the septa are often incomplete in annelids that are semi-sessile or that do not move by peristalsis or by movements of parapodia – for example some move by whipping movements of the body, some small marine species move by means of cilia (fine muscle-powered hairs) and some burrowers turn their pharynges (throats) inside out to penetrate the sea-floor and drag themselves into it.

The fluid in the coelomata contains coelomocyte cells that defend the animals against parasites and infections. In some species coelomocytes may also contain a respiratory pigment – red hemoglobin in some species, green chlorocruorin in others (dissolved in the plasma) – and provide oxygen transport within their segments. Respiratory pigment is also dissolved in the blood plasma. Species with well-developed septa generally also have blood vessels running all long their bodies above and below the gut, the upper one carrying blood forwards while the lower one carries it backwards. Networks of capillaries in the body wall and around the gut transfer blood between the main blood vessels and to parts of the segment that need oxygen and nutrients. Both of the major vessels, especially the upper one, can pump blood by contracting. In some annelids the forward end of the upper blood vessel is enlarged with muscles to form a heart, while in the forward ends of many earthworms some of the vessels that connect the upper and lower main vessels function as hearts. Species with poorly developed or no septa generally have no blood vessels and rely on the circulation within the coelom for delivering nutrients and oxygen.

However, leeches and their closest relatives have a body structure that is very uniform within the group but significantly different from that of other annelids, including other members of the Clitellata. In leeches there are no septa, the connective tissue layer of the body wall is so thick that it occupies much of the body, and the two coelomata are widely separated and run the length of the body. They function as the main blood vessels, although they are side-by-side rather than upper and lower. However, they are lined with mesothelium, like the coelomata and unlike the blood vessels of other annelids. Leeches generally use suckers at their front and rear ends to move like inchworms. The anus is on the upper surface of the pygidium.

Respiration

In some annelids, including earthworms, all respiration is via the skin. However, many polychaetes and some clitellates (the group to which earthworms belong) have gills associated with most segments, often as extensions of the parapodia in polychaetes. The gills of tube-dwellers and burrowers usually cluster around whichever end has the stronger water flow.

Feeding and Excretion

Lamellibrachian tube worms have no gut and gain nutrients from chemoautotrophic bacteria living inside them.

Feeding structures in the mouth region vary widely, and have little correlation with the animals' diets. Many polychaetes have a muscular pharynx that can be everted (turned inside out to extend it). In these animals the foremost few segments often lack septa so that, when the muscles in these segments contract, the sharp increase in fluid pressure from all these segments everts the pharynx very quickly. Two families, the Eunicidae and Phyllodocidae, have evolved jaws, which can be used for seizing prey, biting off pieces of vegetation, or grasping dead and decaying matter. On the other hand, some predatory polychaetes have neither jaws nor eversible pharynges. Selective deposit feeders generally live in tubes on the sea-floor and use palps to find food particles in the sediment and then wipe them into their mouths. Filter feeders use "crowns" of palps covered in cilia that wash food particles towards their mouths. Non-selective deposit feeders ingest soil or marine sediments via mouths that are generally unspecialized. Some clitellates have sticky pads in the roofs of their mouths, and some of these can evert the pads to capture prey. Leeches often have an eversible proboscis, or a muscular pharynx with two or three teeth.

The gut is generally an almost straight tube supported by the mesenteries (vertical partitions within segments), and ends with the anus on the underside of the pygidium. However, in members of the tube-dwelling family Siboglinidae the gut is blocked by a swollen lining that houses symbiotic bacteria, which can make up 15% of the worms' total weight. The bacteria convert inorganic matter – such as hydrogen sulfide and carbon dioxide from hydrothermal vents, or methane from seeps – to organic matter that feeds themselves and their hosts, while the worms extend their palps into the gas flows to absorb the gases needed by the bacteria.

Annelids with blood vessels use metanephridia to remove soluble waste products, while those without use protonephridia. Both of these systems use a two-stage filtration process, in which fluid and waste products are first extracted and these are filtered again to re-absorb any re-usable materials while dumping toxic and spent materials as urine. The difference is that protonephridia combine both filtration stages in the same organ, while metanephridia perform only the second filtration and rely on other mechanisms for the first – in annelids special filter cells in the walls of the blood vessels let fluids and other small molecules pass into the coelomic fluid, where it circulates to the metanephridia. In annelids the points at which fluid enters the protonephridia or metanephridia are on the forward side of a septum while the second-stage filter and the nephridiopore (exit opening in the body wall) are in the following segment. As a result, the hindmost segment (before the growth zone and pygidium) has no structure that extracts its wastes, as there is no following segment to filter and discharge them, while the first segment contains an extraction structure that passes wastes to the second, but does not contain the structures that re-filter and discharge urine.

Reproduction and Life Cycle

Asexual Reproduction

This sabellid tubeworm is budding

Polychaetes can reproduce asexually, by dividing into two or more pieces or by budding off a new individual while the parent remains a complete organism. Some oligochaetes, such as *Aulophorus furcatus*, seem to reproduce entirely asexually, while others reproduce asexually in summer and sexually in autumn. Asexual reproduction in oligochaetes is always by dividing into two or more pieces, rather than by budding. However, leeches have never been seen reproducing asexually.

Most polychaetes and oligochaetes also use similar mechanisms to regenerate after suffering damage. Two polychaete genera, *Chaetopterus* and *Dodecaceria*, can regenerate from a single segment, and others can regenerate even if their heads are removed. Annelids are the most complex animals that can regenerate after such severe damage. On the other hand, leeches cannot regenerate.

Sexual Reproduction

It is thought that annelids were originally animals with two separate sexes, which released ova and sperm into the water via their nephridia. The fertilized eggs develop into trochophore larvae, which live as plankton. Later they sink to the sea-floor and metamorphose into miniature adults: the part of the trochophore between the apical tuft and the prototroch becomes the prostomium (head); a small area round the trochophore's anus becomes the pygidium (tail-piece); a narrow band immediately in front of that becomes the growth zone that produces new segments; and the rest of the trochophore becomes the peristomium (the segment that contains the mouth).

However, the lifecycles of most living polychaetes, which are almost all marine animals, are unknown, and only about 25% of the 300+ species whose lifecycles are known follow this pattern. About 14% use a similar external fertilization but produce yolk-rich eggs, which reduce the time the larva needs to spend among the plankton, or eggs from which miniature adults emerge rather than larvae. The rest care for the fertilized eggs until they hatch – some by producing jelly-covered masses of eggs which they tend, some by attaching the eggs to their bodies and a few species by keeping the eggs within their bodies until they hatch. These species use a variety of methods for sperm transfer; for example, in some the females collect sperm released into the water, while in others the males have a penis that inject sperm into the female. There is no guarantee that this is a representative sample of polychaetes' reproductive patterns, and it simply reflects scientists' current knowledge.

Some polychaetes breed only once in their lives, while others breed almost continuously or through several breeding seasons. While most polychaetes remain of one sex all their lives, a significant percentage of species are full hermaphrodites or change sex during their lives. Most polychaetes whose reproduction has been studied lack permanent gonads, and it is uncertain how they produce ova and sperm. In a few species the rear of the body splits off and becomes a separate individual that lives just long enough to swim to a suitable environment, usually near the surface, and spawn.

Most mature clitellates (the group that includes earthworms and leeches) are full hermaphrodites, although in a few leech species younger adults function as males and become female at maturity. All have well-developed gonads, and all copulate. Earthworms store their partners' sperm in spermathecae ("sperm stores") and then the clitellum produces a cocoon that collects ova from the ovaries and then sperm from the spermathecae. Fertilization and development of earthworm eggs takes place in the cocoon. Leeches' eggs are fertilized in the ovaries, and then transferred to the cocoon. In all clitellates the cocoon also either produces yolk when the eggs are fertilized or nutrients while they are developing. All clitellates hatch as miniature adults rather than larvae.

Ecological Significance

Charles Darwin's book *The Formation of Vegetable Mould through the Action of Worms* (1881) presented the first scientific analysis of earthworms' contributions to soil fertility. Some burrow while others live entirely on the surface, generally in moist leaf litter. The burrowers loosen the soil so that oxygen and water can penetrate it, and both surface and burrowing worms help to produce soil by mixing organic and mineral matter, by accelerating the decomposition of organic matter and thus making it more quickly available to other organisms, and by concentrating minerals and converting them to forms that plants can use more easily. Earthworms are also important prey for birds ranging in size from robins to storks, and for mammals ranging from shrews to badgers, and in some cases conserving earthworms may be essential for conserving endangered birds.

Terrestrial annelids can be invasive in some situations. In the glaciated areas of North America, for example, almost all native earthworms are thought to have been killed by the glaciers and the worms currently found in those areas are all introduced from other areas, primarily from Europe, and, more recently, from Asia. Northern hardwood forests are especially negatively impacted by invasive worms through the loss of leaf duff, soil fertility, changes in soil chemistry and the loss of ecological diversity. Especially of concern is *Amynthas agrestis* and at least one state (Wisconsin) has listed it as a prohibited species.

Earthworms migrate only a limited distance annually on their own, and the spread of invasive worms is increased rapidly by anglers and from worms or their cocoons in the dirt on vehicle tires or footwear.

Marine annelids may account for over one-third of bottom-dwelling animal species around coral reefs and in tidal zones. Burrowing species increase the penetration of water and oxygen into the sea-floor sediment, which encourages the growth of populations of aerobic bacteria and small animals alongside their burrows.

Although blood-sucking leeches do little direct harm to their victims, some transmit flagellates that can be very dangerous to their hosts. Some small tube-dwelling oligochaetes transmit myxosporean parasites that cause whirling disease in fish.

Interaction with Humans

Earthworms make a significant contribution to soil fertility. The rear end of the Palolo worm, a marine polychaete that tunnels through coral, detaches in order to spawn at the surface, and the people of Samoa regard these spawning modules as a delicacy. Anglers sometimes find that worms are more effective bait than artificial flies, and worms can be kept for several days in a tin lined with damp moss. Ragworms are commercially important as bait and as food sources for aquaculture, and there have been proposals to farm them in order to reduce over-fishing of their natural populations. Some marine polychaetes' predation on molluscs causes serious losses to fishery and aquaculture operations.

Scientists study aquatic annelids to monitor the oxygen content, salinity and pollution levels in fresh and marine water.

Accounts of the use of leeches for the medically dubious practise of blood-letting have come from China around 30 AD, India around 200 AD, ancient Rome around 50 AD and later throughout Europe. In the 19th century medical demand for leeches was so high that some areas' stocks were exhausted and other regions imposed restrictions or bans on exports, and *Hirudo medicinalis* is treated as an endangered species by both IUCN and CITES. More recently leeches have been used to assist in microsurgery, and their saliva has provided anti-inflammatory compounds and several important anticoagulants, one of which also prevents tumors from spreading.

Ragworms' jaws are strong but much lighter than the hard parts of many other organisms, which are biomineralized with calcium salts. These advantages have attracted the attention of engineers. Investigations showed that ragworm jaws are made of unusual proteins that bind strongly to zinc.

Evolutionary History

Fossil Record

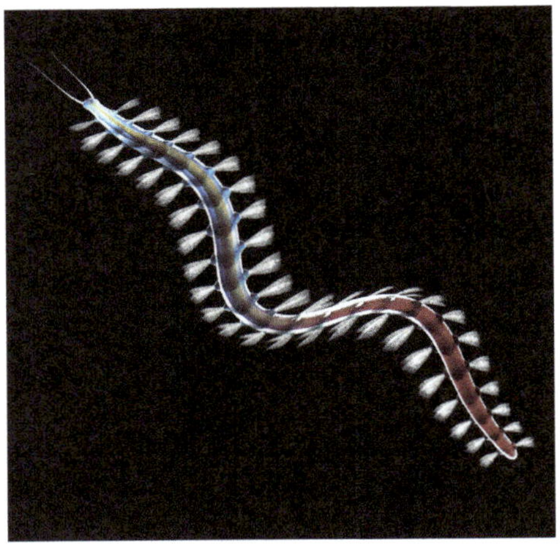

Burgessochaeta setigera

Since annelids are soft-bodied, their fossils are rare. Polychaetes' fossil record consists mainly of the jaws that some species had and the mineralized tubes that some secreted. Some Ediacaran fossils such as *Dickinsonia* in some ways resemble polychaetes, but the similarities are too vague for these fossils to be classified with confidence. The small shelly fossil *Cloudina*, from 549 to 542 million years ago, has been classified by some authors as an annelid, but by others as a cnidarian (i.e. in the phylum to which jellyfish and sea anemones belong). Until 2008 the earliest fossils widely accepted as annelids were the polychaetes *Canadia* and *Burgessochaeta*, both from Canada's Burgess Shale, formed about 505 million years ago in the early Cambrian. *Myoscolex*, found in Australia and a little older than the Burgess Shale, was possibly an annelid. However, it lacks some typical annelid features and has features which are not usually found in annelids and some of which are associated with other phyla. Then Simon Conway Morris and John Peel reported *Phragmochaeta* from Sirius Passet, about 518 million years old, and concluded that it was the oldest annelid known to date. There has been vigorous debate about whether the Burgess Shale fossil *Wiwaxia* was a mollusc or an annelid. Polychaetes diversified in the early Ordovician, about 488 to 474 million years ago. It is not until the early Ordovician that the first annelid jaws are found, thus the crown-group cannot have appeared before this date and probably appeared somewhat later. By the end of the Carboniferous, about 299 million years ago, fossils of most of the modern mobile polychaete groups had appeared. Many fossil tubes look like those made by modern sessile polychaetes , but the first tubes clearly produced by polychaetes date from the Jurassic, less than 199 million years ago.

The earliest good evidence for oligochaetes occurs in the Tertiary period, which began 65 million years ago, and it has been suggested that these animals evolved around the same time as flowering plants in the early Cretaceous, from 130 to 90 million years ago. A trace fossil consisting of a convoluted burrow partly filled with small fecal pellets may be evidence that earthworms were present in the early Triassic period from 251 to 245 million years ago. Body fossils going back to the mid Ordovician, from 472 to 461 million years ago, have been tentatively classified as oligochaetes, but these identifications are uncertain and some have been disputed.

Family Tree

Traditionally the annelids have been divided into two major groups, the polychaetes and clitellates. In turn the clitellates were divided into oligochaetes, which include earthworms, and hirudinomorphs, whose best-known members are leeches. For many years there was no clear arrangement of the approximately 80 polychaete families into higher-level groups. In 1997 Greg Rouse and Kristian Fauchald attempted a "first heuristic step in terms of bringing polychaete systematics to an acceptable level of rigour", based on anatomical structures, and divided polychaetes into:

- Scolecida, less than 1,000 burrowing species that look rather like earthworms.
- Palpata, the great majority of polychaetes, divided into:
    - Canalipalpata, which are distinguished by having long grooved palps that they use for feeding, and most of which live in tubes.
    - Aciculata, the most active polychaetes, which have parapodia reinforced by internal spines (aciculae).

Also in 1997 Damhnait McHugh, using molecular phylogenetics to compare similarities and differences in one gene, presented a very different view, in which: the clitellates were an offshoot of one branch of the polychaete family tree; the pogonophorans and echiurans, which for a few decades had been regarded as a separate phyla, were placed on other branches of the polychaete tree. Subsequent molecular phylogenetics analyses on a similar scale presented similar conclusions.

In 2007 Torsten Struck and colleagues compared 3 genes in 81 taxa, of which 9 were outgroups, in other words not considered closely related to annelids but included to give an indication of where the organisms under study are placed on the larger tree of life. For a cross-check the study used an analysis of 11 genes (including the original 3) in 10 taxa. This analysis agreed that clitellates, pogonophorans and echiurans were on various branches of the polychaete family tree. It also concluded that the classification of polychaetes into Scolecida, Canalipalpata and Aciculata was useless, as the members of these alleged groups were scattered all over the family tree derived from comparing the 81 taxa. In addition, it also placed sipunculans, generally regarded at the time as a separate phylum, on another branch of the polychaete tree, and concluded that leeches were a sub-group of oligochaetes rather than their sister-group among the clitellates. Rouse accepted the analyses based on molecular phylogenetics, and their main conclusions are now the scientific consensus, although the details of the annelid family tree remain uncertain.

In addition to re-writing the classification of annelids and 3 previously independent phyla, the molecular phylogenetics analyses undermine the emphasis that decades of previous writings placed on the importance of segmentation in the classification of invertebrates. Polychaetes, which these analyses found to be the parent group, have completely segmented bodies, while polychaetes' echiurans and sipunculan offshoots are not segmented and pogonophores are segmented only in the rear parts of their bodies. It now seems that segmentation can appear and disappear much more easily in the course of evolution than was previously thought. The 2007 study also noted that the ladder-like nervous system, which is associated with segmentation, is less universal previously thought in both annelids and arthropods.

Annelids are members of the protostomes, one of the two major superphyla of bilaterian animals – the other is the deuterostomes, which includes vertebrates. Within the protostomes, annelids used to be grouped with arthropods under the super-group Articulata ("jointed animals"), as segmentation is obvious in most members of both phyla. However, the genes that drive segmentation in arthropods do not appear to do the same in annelids. Arthropods and annelids both have close relatives that are unsegmented. It is at least as easy to assume that they evolved segmented bodies independently as it is to assume that the ancestral protostome or bilaterian was segmented and that segmentation disappeared in many descendant phyla. The current view is that annelids are grouped with molluscs, brachiopods and several other phyla that have lophophores (fan-like feeding structures) and/or trochophore larvae as members of Lophotrochozoa. Bryozoa may be the most basal phylum (the one that first became distinctive) within the Lophotrochozoa, and the relationships between the other members are not yet known. Arthropods are now regarded as members of the Ecdysozoa ("animals that molt"), along with some phyla that are unsegmented.

The "Lophotrochozoa" hypothesis is also supported by the fact that many phyla within this group, including annelids, molluscs, nemerteans and flatworms, follow a similar pattern in the fertilized egg's development. When their cells divide after the 4-cell stage, descendants of these 4 cells form a spiral pattern. In these phyla the "fates" of the embryo's cells, in other words the roles their descendants will play in the adult animal, are the same and can be predicted from a very early stage. Hence this development pattern is often described as "spiral determinate cleavage".

Brachiopod

Brachiopods, phylum Brachiopoda, are a group of lophotrochozoan animals that have hard "valves" (shells) on the upper and lower surfaces, unlike the left and right arrangement in bivalve molluscs. Brachiopod valves are hinged at the rear end, while the front can be opened for feeding or closed for protection. Two major groups are recognized, articulate and inarticulate. Articulate brachiopods have toothed hinges and simple opening and closing muscles, while inarticulate brachiopods have untoothed hinges and a more complex system of muscles used to keep the two halves aligned. In a typical brachiopod a stalk-like pedicle projects from an opening in one of the valves near the hinges, known as the pedicle valve, keeping the animal anchored to the seabed but clear of silt that would obstruct the opening.

The word "brachiopod" is formed from the Ancient Greek words ("arm") and ("foot"). They are often known as "lamp shells", since the curved shells of the class Terebratulida look rather like pottery oil-lamps.

Lifespans range from three to over thirty years. Ripe gametes (ova or sperm) float from the gonads into the main coelom and then exit into the mantle cavity. The larvae of inarticulate brachiopods are miniature adults, with lophophores that enable the larvae to feed and swim for months until the animals become heavy enough to settle to the seabed. The planktonic larvae of articulate species do not resemble the adults, but rather look like blobs with yolk sacs, and remain among the plankton for only a few days before leaving the water column upon metamorphosing.

In addition to the traditional classification of brachiopods into inarticulate and articulate, two approaches appeared in the 1990s: one approach groups the inarticulate Craniida with articulate brachiopods, since both use the same material in the mineral layers of their shell; the other approach makes the Craniida a third group, as their outer organic layer is different from that in either of the other two. However, some taxonomists believe it is premature to suggest higher levels of classification such as order and recommend a bottom-up approach that identifies genera and then groups these into intermediate groups. Traditionally, brachiopods have been regarded as members of, or as a sister group to, the deuterostomes, a superphylum that includes chordates and echinoderms. One type of analysis of the evolutionary relationships of brachiopods has always placed brachiopods as protostomes while another type has split between placing brachiopods among the protostomes or the deuterostomes.

It was suggested in 2003 that brachiopods had evolved from an ancestor similar to *Halkieria*, a slug-like Cambrian animal with "chain mail" on its back and a shell at the front and rear end; it was thought that the ancestral brachiopod converted its shells into a pair of valves by folding the rear part of its body under its front. However, new fossils found in 2007 and 2008 showed that the "chain mail" of tommotiids formed the tube of a sessile animal; one tommotiid resembled phoronids, which are close relatives or a subgroup of brachiopods, while the other tommotiid bore two symmetrical plates that might be an early form of brachiopod valves. Lineages of brachiopods that have both fossil and extant taxa appeared in the early Cambrian, Ordovician, and Carboniferous periods, respectively. Other lineages have arisen and then become extinct, sometimes during severe mass extinctions. At their peak in the Paleozoic era, the brachiopods were among the most abundant filter-feeders and reef-builders, and occupied other ecological niches, including swimming in the jet-propulsion style of scallops. Brachiopod fossils have been useful indicators of climate changes during the Paleozoic. However, after the Permian–Triassic extinction event, brachiopods recovered only a third of their former diversity. A study in 2007 concluded the brachiopods were especially vulnerable to the Permian–Triassic extinction, as they built calcareous hard parts (made of calcium carbonate) and had low metabolic rates and weak respiratory systems. It was often thought that brachiopods went into decline after the Permian–Triassic extinction, and were out-competed by bivalves, but a study in 1980 found both brachiopod and bivalve species increased from the Paleozoic to modern times, with bivalves increasing faster; after the Permian–Triassic extinction, brachiopods became for the first time less diverse than bivalves.

Brachiopods live only in the sea, and most species avoid locations with strong currents or waves. The larvae of articulate species settle in quickly and form dense populations in well-defined areas while the larvae of inarticulate species swim for up to a month and have wide ranges. Brachiopods now live mainly in cold water and low light. Fish and crustaceans seem to find brachiopod flesh distasteful and seldom attack them. Among brachiopods, only the lingulids (*Lingula sp.*) have been fished commercially, on a very small scale. One brachiopod species (*Coptothyrus adamsi*) may be a measure of environmental conditions around an oil terminal being built in Russia on the shore of the Sea of Japan.

Description

Shells and their Mechanisms

Modern brachiopods range from 1 to 100 millimetres (0.039 to 3.937 in) long, and most species are about 10 to 30 millimetres (0.39 to 1.18 in). The largest brachiopods known – *Gigantoproductus* and *Titanaria*, reaching 30 to 38 centimetres (12 to 15 in) in width – occurred in the upper part of the Lower Carboniferous. Each has two valves (shell sections) which cover the dorsal and ventral surface of the animal, unlike bivalve molluscs whose shells cover the lateral surfaces. The valves are termed brachial and pedicle. The brachial valve bears on its inner surface the brachia ("arms") from which the phylum gets its name, and which support the lophophore, used for feeding and respiration. The pedicle valve has on its inner surface the attachment to the stalk-like pedicle by which most brachiopods attach themselves to the substrate. (R. C. Moore, 1952) The brachial and pedicle valves are often called the dorsal ("upper") and ventral ("lower"), but some paleontologists regard the terms "dorsal" and "ventral" as irrelevant since they believe that the "ventral" valve was formed by a folding of the upper surface under the body. The pedicle valve is typically larger than the brachial. In most articulate brachiopod species, both valves are convex, the surfaces often bearing growth lines and/or other ornamentation. However, inarticulate lingulids, which burrow into the seabed, have valves that are smoother, flatter and of similar size and shape. (R. C. Moore, 1952)

Articulate ("jointed") brachiopods have a tooth and socket arrangement by which the pedicle and brachial valves hinge, locking the valves against lateral displacement. Inarticulate brachiopods have no matching teeth and sockets; their valves are held together only by muscles. (R. C. Moore, 1952)

The inarticulate species *Lingula anatina*, showing the long pedicle, flattened shells and prominent chaetae around the front edge of the shells

All brachiopods have adductor muscles that are set on the inside of the pedicle valve and which close the valves by pulling on the part of the brachial valve ahead of the hinge. These muscles have both "quick" fibers that close the valves in emergencies and "catch" fibers that are slower but can keep the valves closed for long periods. Articulate brachiopods open the valves by means of abductor muscles, also known as diductors, which lie further to the rear and pull on the part of the brachial valve behind the hinge. Inarticulate brachiopods use a different opening mechanism, in which muscles reduce the length of the coelom (main body cavity) and make it bulge outwards, pushing the valves apart. Both classes open the valves to an angle of about 10°. The more complex set of muscles employed by inarticulate brachiopods can also operate the valves as scissors, a mechanism that lingulids use to burrow.

Each valve consists of three layers, an outer periostracum made of organic compounds and two biomineralized layers. Articulated brachiopods have an outermost periostracum made of proteins, a "primary layer" of calcite (a form of calcium carbonate) under that, and innermost a mixture of proteins and calcite. Inarticulate brachiopod shells have a similar sequence of layers, but their composition is different from that of articulated brachiopods and also varies among the classes of inarticulate brachiopods. Lingulids and discinids, which have pedicles, have a matrix of glycosaminoglycans (long, unbranched polysaccharides), in which other materials are embedded: chitin in the periostracum; apatite containing calcium phosphate in the primary biomineralized layer; and a complex mixture in the innermost layer, containing collagen and other proteins, chitinophosphate and apatite. Craniids, which have no pedicle and cement themselves directly to hard surfaces, have a periostracum of chitin and mineralized layers of calcite. Shells can grow holoperipherally, where new material is added all around the margin, or mixoperipherally, where new material is added to the posterior region of the shell in an anterior direction, growing towards the other shell.

Mantle

Brachiopods, as with molluscs, have an epithelial mantle that secretes and lines the shell, and encloses the internal organs. The brachiopod body occupies only about one-third of the internal space inside the shell, nearest the hinge. The rest of the space is lined with the mantle lobes, extensions that enclose a water-filled space in which sits the lophophore. The coelom extends into each lobe as a network of canals, which carry nutrients to the edges of the mantle.

Relatively new cells in a groove on the edges of the mantle secrete material that extends the periostracum. These cells are gradually displaced to the underside of the mantle by more recent cells in the groove, and switch to secreting the mineralized material of the shell valves. In other words, on the edge of the valve the periostracum is extended first, and then reinforced by extension of the mineralized layers under the periostracum. In most species the edge of the mantle also bears movable bristles, often called chaetae or setae, that may help defend the animals and may act as sensors. In some brachiopods groups of chaetae help to channel the flow of water into and out of the mantle cavity.

In most brachiopods, diverticula (hollow extensions) of the mantle penetrate through the mineralized layers of the valves into the periostraca. The function of these diverticula is uncertain and it is suggested that they may be storage chambers for chemicals such as glycogen, may secrete repellents to deter organisms that stick to the shell or may help in respiration. Experiments show that a brachiopod's oxygen consumption drops if petroleum jelly is smeared on the shell, clogging the diverticula.

Lophophore

Like bryozoans and phoronids, brachiopods have a lophophore, a crown of tentacles whose cilia (fine hairs) create a water current that enables them to filter food particles out of the water. However a bryozoan or phoronid lophophore is a ring of tentacles mounted on a single, retracted stalk, while the basic form of the brachiopod lophophore is U-shaped, forming the brachia ("arms") from which the phylum gets its name. Brachiopod lophophores are non-retractable and occupy up to two-thirds of the internal space, in the frontmost area where the valves gape when opened. To provide enough filtering capacity in this restricted space, lophophores of larger brachiopods are folded in moderately to very complex shapes—loops and coils are common, and some species' lophophores resemble a hand with the fingers splayed. In all species the lophophore is supported by cartilage and by a hydrostatic skeleton (in other words by the pressure of its internal fluid), and the fluid extends into the tentacles. Some articulate brachiopods also have a brachidium, a calcareous support for the lophophore attached to the inside of the brachial valve.

The tentacles bear cilia (fine mobile hairs) on their edges and along the center. The beating of the outer cilia drives a water current from the tips of the tentacles to their bases, where it exits. Food particles that collide with the tentacles are trapped by mucus, and the cillia down the middle drive this mixture to the base of the tentacles. A brachial groove runs round the bases of the tentacles, and its own cilia pass food along the groove towards the mouth. The method used by brachiopods is known as "upstream collecting", as food particles are captured as they enter the field of cilia that creates the feeding current. This method is used by the related phoronids and bryozoans, and also by pterobranchs. Entoprocts use a similar-looking crown of tentacles, but it is solid and the flow runs from bases to tips, forming a "downstream collecting" system that catches food particles as they are about to exit.

Attachment to Substrate

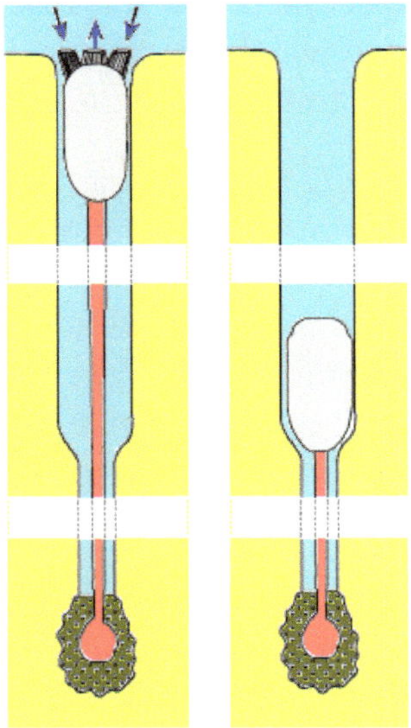

A lingulid in its burrow, in "up" and retracted positions

Most modern species attach to hard surfaces by means of a cylindrical pedicle ("stalk"), an extension of the body wall. This has a chitinous cuticle (non-cellular "skin") and protrudes through an opening in the hinge. However, some genera such as the inarticulate *Crania* and the articulate *Lacazella* have no pedicle, and cement the rear of the "pedicle" valve to a surface so that the front is slightly inclined up away from the surface. In a few articulate genera such as *Neothyris* and *Anakinetica*, the pedicles wither as the adults grow and finally lie loosely on the surface. In these genera the shells are thickened and shaped so that the opening of the gaping valves is kept free of the sediment.

Pedicles of inarticulate species are extensions of the main coelom, which houses the internal organs. A layer of longitudinal muscles lines the epidermis of the pedicle. Members of the order

Lingulida have long pedicles, which they use to burrow into soft substrates, to raise the shell to the opening of the burrow to feed, and to retract the shell when disturbed. A lingulid moves its body up and down the top two-thirds of the burrow, while the remaining third is occupied only by the pedicle, with a bulb on the end that builds a "concrete" anchor. However, the pedicles of the order Discinida are short and attach to hard surfaces.

An articulate pedicle has no coelom, is constructed from a different part of the larval body, and has a core composed of connective tissue. Muscles at the rear of the body can straighten, bend or even rotate the pedicle. The far end of the pedicle generally has rootlike extensions or short papillae ("bumps"), which attach to hard surfaces. However, articulate brachiopods of genus *Chlidonophora* use a branched pedicle to anchor in sediment. The pedicle emerges from the pedicle valve, either through a notch in the hinge or, in species where the pedicle valve is longer than the brachial, from a hole where the pedicle valve doubles back to touch the brachial valve. Some species stand with the front end upwards, while others lie horizontal with the pedicle valve uppermost.

Some brachiopods do not have a functional pedicle, and thus do not have this pedicle opening.

Feeding and Excretion

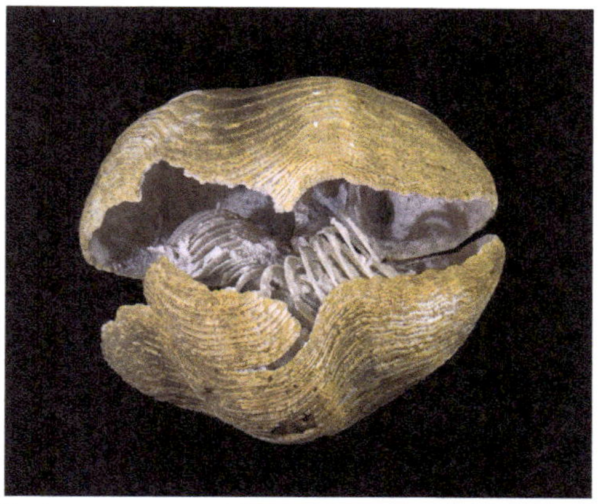

A fossil of *Spiriferina rostrata* with visible brachidium (lophophore support)

Rhynchonellid brachiopod with interior spondylium ("C" in image) visible; Roadian, Guadalupian (Middle Permian); Glass Mountains, Texas.

The water flow enters the lophophore from the sides of the open valves and exits at the front of the animal. In lingulids the entrance and exit channels are formed by groups of chaetae that function as funnels. In other brachiopods the entry and exit channels are organized by the shape of the lophophore. The lophophore captures food particles, especially phytoplankton (tiny photosynthetic organisms), and deliver them to the mouth via the brachial grooves along the bases of the tentacles. The mouth is at the base of the lophophore. Food passes through the mouth, muscular pharynx ("throat") and oesophagus ("gullet"), all of which are lined with cilia and cells that secrete mucus and digestive enzymes. The stomach wall has branched ceca ("pouches") where food is digested, mainly within the cells.

Nutrients are transported throughout the coelom, including the mantle lobes, by cilia. The wastes produced by metabolism are broken into ammonia, which is eliminated by diffusion through the mantle and lophophore. Brachiopods have metanephridia, used by many phyla to excrete ammonia and other dissolved wastes. However, brachiopods have no sign of the podocytes, which perform the first phase of excretion in this process, and brachiopod metanephridia appear to be used only to emit sperm and ova.

The majority of food consumed by brachiopods is digestible, with very little solid waste produced. The cilia of the lophophore can change direction to eject isolated particles of indigestible matter. If the animal encounters larger lumps of undesired matter, the cilia lining the entry channels pause and the tentacles in contact with the lumps move apart to form large gaps and then slowly use their cilia to dump the lumps onto the lining of the mantle. This has its own cilia, which wash the lumps out through the opening between the valves. If the lophophore is clogged, the adductors snap the valves sharply, which creates a "sneeze" that clears the obstructions. In some inarticulate brachiopods the digestive tract is U-shaped and ends with an anus that eliminates solids from the front of the body wall. Other inarticulate brachiopods and all articulate brachiopods have a curved gut that ends blindly, with no anus. These animals bundle solid waste with mucus and periodically "sneeze" it out, using sharp contractions of the gut muscles.

Circulation and Respiration

The lophophore and mantle are the only surfaces that absorb oxygen and eliminate carbon dioxide. Oxygen seems to be distributed by the fluid of the coelom, which is circulated through the mantle and driven either by contractions of the lining of the coelom or by beating of its cilia. In some species oxygen is partly carried by the respiratory pigment hemerythrin, which is transported in coelomocyte cells.The maximum oxygen consumption of brachiopods is low, and their minimum requirement is not measurable.

Brachiopods also have colorless blood, circulated by a muscular heart lying in the dorsal part of the body above the stomach. The blood passes through vessels that extend to the front and back of the body, and branch to organs including the lophophore at the front and the gut, muscles, gonads and nephridia at the rear. The blood circulation seems not to be completely closed, and the coelomic fluid and blood must mix to a degree. The main function of the blood may be to deliver nutrients.

Nervous System and Senses

The "brain" of adult articulates consists of two ganglia, one above and the other below the oesophagus. Adult inarticulates have only the lower ganglion. From the ganglia and the commissures where they join, nerves run to the lophophore, the mantle lobes and the muscles that operate the valves. The edge of the mantle has probably the greatest concentration of sensors. Although not directly connected to sensory neurons, the mantle's chaetae probably send tactile signals to receptors in the epidermis of the mantle. Many brachiopods close their valves if shadows appear above them, but the cells responsible for this are unknown. Some brachiopods have statocysts, which detect changes in the animals' position.

Reproduction and Life Cycle

Lifespans range from 3 to over 30 years. Adults of most species are of one sex throughout their lives. The gonads are masses of developing gametes (ova or sperm), and most species have four gonads, two in each valve. Those of articulates lie in the channels of the mantle lobes, while those of inarticulates lie near the gut. Ripe gametes float into the main coelom and then exit into the mantle cavity via the metanephridia, which open on either side of the mouth. Most species release both ova and sperm into the water, but females of some species keep the embryos in brood chambers until the larvae hatch.

The cell division in the embryo is radial (cells form in stacks of rings directly above each other), holoblastic (cells are separate, although adjoining) and regulative (the type of tissue into which a cell develops is controlled by interactions between adjacent cells, rather than rigidly within each cell). While some animals develop the mouth and anus by deepening the blastopore, a "dent" in the surface of the early embryo, the blastopore of brachiopods closes up, and their mouth and anus develop from new openings.

The larvae of inarticulates swim as plankton for months and are like miniature adults, with valves, mantle lobes, a pedicle that coils in the mantle cavity, and a small lophophore, which is used for both feeding and swimming—except that Craniids have no pedicle. As the shell becomes heavier, the juvenile sinks to the bottom and becomes a sessile adult. The larvae of articulate species live only on yolk, and remain among the plankton for only a few days. This type of larva has a ciliated frontmost lobe that becomes the body and lophophore, a rear lobe that becomes the pedicle, and a mantle like a skirt, with the hem towards the rear. On metamorphosing into an adult, the pedicle attaches to a surface, the front lobe develops the lophophore and other organs, and the mantle rolls up over the front lobe and starts to secrete the shell. In cold seas, brachiopod growth is seasonal and the animals often lose weight in winter. These variations in growth often form growth lines in the shells. Members of some genera have survived for a year in aquaria without food.

Taxonomy

Taxonomical History

Brachiopod fossils show great diversity in the morphology of the shells and lophophore, while the modern genera show less diversity but provide soft-bodied characteristics. Both fossils and extant species have limitations that make it difficult to produce a comprehensive classification of brachiopods based on morphology. The phylum also has experienced significant convergent evolution and reversals (in which a more recent group seems to have lost a characteristic that is seen in an intermediate group, reverting to a characteristic last seen in an older group). Hence some brachiopod taxonomists believe it is premature to define higher levels of classification such as order, and recommend instead a bottom-up approach that identifies genera and then groups these into intermediate groups.

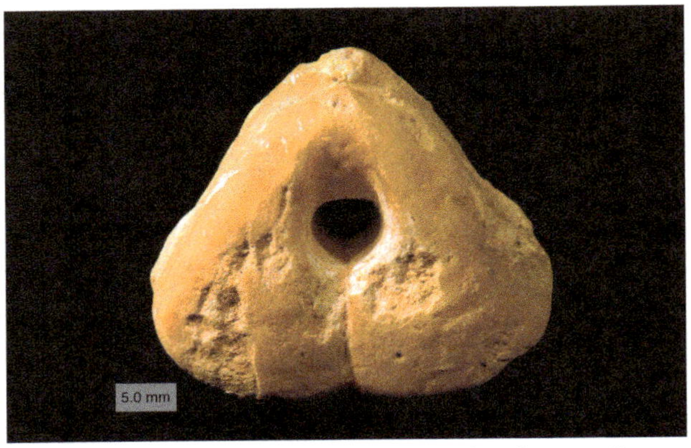

Pygites diphyoides (d'Orbigny, 1849) from the Hauterivian (Lower Cretaceous) of Cehegin, Murcia, Spain. This terebratulid is characterized by a central perforation through its valves.

However, other taxonomists believe that some patterns of characteristics are sufficiently stable to make higher-level classifications worthwhile, although there are different views about what the higher-level classifications should be. The "traditional" classification was defined in 1869; two further approaches were established in the 1990s:

- In the "traditional" classification, the Articulata have toothed hinges between the valves, while the hinges of the Inarticulata are held together only by muscles.
- A classification devised in the 1990s, based on the materials of which the shells are based, united the Craniida and the "articulate" brachiopods in the Calciata, which have calcite shells. The Lingulida and Discinida, combined in the Lingulata, have shells made of chitin and calcium phosphate.
- A three-part scheme, also from the 1990s, places the Craniida in a separate group of its own, the Craniformea. The Lingulida and Discinida are grouped as Linguliformea, and the Rhynchonellida and Terebratulida as Rhynchonelliformea.

About 330 living species are recognized, grouped into over 100 genera. The great majority of modern brachiopods are rhynchonelliforms (Articulata, but excluding Craniida).

Modern Classification

Genetic analysis performed since the 1990s has extended the understanding of the relationship between different organisms. It is now clear the brachiopods do not belong to the Deuterostomata (such as echinoderms and chordates) as was hypothesized earlier, but should be included in the Protostomia (with mollusks and annelid worms), in a subgroup now called Lophotrochozoa. Although their adult morphology seems rather different, the nucleotid sequence of the 18S rRNA indicates that the horseshoe worms are the closest relatives of the inarticulate brachiopods rather than articulate brachiopods. For now, the weight of evidence is inconclusive as to the exact relations within the inarticulates. Consequently, it has been suggested to included horseshoe worms in the Brachiopoda as a class named Phoronata B.L.Cohen&Weydmann in addition to the Craniata and Lingulata, within the subphylum Linguliformea. The other subphylum Rhynchonelliformea only contains one class, which is subdivided in the orders Rhynchonellida, Terebratulida and Thecideida.

Ecology

Strophomenid brachiopod with attached cornulitid worm tube (Upper Ordovician, SE Indiana, USA). Brachiopod valves often serve as substrates for encrusting organisms.

Distribution and Habitat

Brachiopods live only in the sea. Most species avoid locations with strong currents or waves, and typical sites include rocky overhangs, crevices and caves, steep slopes of continental shelves, and in deep ocean floors. However, some articulate species attach to kelp or in exceptionally sheltered sites in intertidal zones. The smallest living brachiopod, *Gwynia*, is only about 1 millimetre (0.039 in) long, and lives in between gravel. Rhynchonelliforms (Articulata excluding Craniida), whose larvae consume only their yolks and settle and develop quickly, specialize in specific areas and form dense populations that can reach thousands per meter. Young adults often attach to the shells of more mature ones. On the other hand, inarticulate brachiopods, whose larva swim for up to a month before settling, have wide ranges. Members of the discinoid genus *Pelagodiscus* have a cosmopolitan distribution.

Interactions with Other Organisms

The rates of metabolism of Brachiopoda are between one third and one tenth of those of bivalves. While brachiopods were abundant in warm, shallow seas during the Cretaceous period, they have been outcompeted by bivalves, and now live mainly in cold and low-light conditions.

Brachiopod shells occasionally show evidence of damage by predators, and sometimes of subsequent repair. Fish and crustaceans seem to find brachiopod flesh distasteful. The fossil record shows that drilling predators like gastropods attacked molluscs and echinoids 10 to 20 times more often than they did brachiopods, suggesting that such predators attacked brachiopods by mistake or when other prey was scarce. In waters where food is scarce, the snail *Capulus ungaricus* steals food from bivalves, snails, tube worms, and brachiopods.

Among brachiopods only the lingulids have been fished commercially, and only on a very small scale. Brachiopods seldom settle on artificial surfaces, probably because they are vulnerable to pollution. This may make the population of *Coptothyrus adamsi* useful as a measure of environmental conditions around an oil terminal being built in Russia on the shore of the Sea of Japan.

Evolutionary History

Isocrania costata, Upper Maastrichtian (Upper Cretaceous), Maastricht, The Netherlands

Fossil Record

Over 12,000 fossil species are recognized,grouped into over 5,000 genera. While the largest modern brachiopods are 100 millimetres (3.9 in) long, a few fossils measure up to 200 millimetres (7.9 in) wide. The earliest confirmed brachiopods have been found in the early Cambrian, inarticulate forms appearing first, followed soon after by articulate forms. Three unmineralized species have also been found in the Cambrian, and apparently represent two distinct groups that evolved from mineralized ancestors. The inarticulate *Lingula* is often called a "living fossil", as very similar genera have been found all the way back to the Ordovician. On the other hand, articulate brachiopods have produced major diversifications, and suffered severe mass extinctions—but the articulate Rhynchonellida and Terebratulida, the most diverse present-day groups, appeared at the start of the Ordovician and Carboniferous respectively.

Since 1991 Claus Nielsen has proposed a hypothesis about the development of brachiopods, adapted in 2003 by Cohen and colleagues as a hypothesis about the earliest evolution of brachiopods. This "brachiopod fold" hypothesis suggests that brachiopods evolved from an ancestor similar to *Halkieria*, a slug-like animal with "chain mail" on its back and a shell at the front and rear end. The hypothesis proposes that the first brachiopod converted its shells into a pair of valves by folding the rear part of its body under its front.

However, fossils from 2007 onwards have supported a new interpretation of the Early-Cambrian tommotiids, and a new hypothesis that brachiopods evolved from tommotiids. The "armor mail" of tommotiids was well-known but not in an assembled form, and it was generally assumed that tommotiids were slug-like animals similar to *Halkieria*, except that tommotiids' armor was made of organophosphatic compounds while that of *Halkieria* was made of calcite. However, fossils of a new tommotiid, *Eccentrotheca*, showed an assembled mail coat that formed a tube, which would indicate a sessile animal rather than a creeping slug-like one. *Eccentrotheca*'s organophosphatic tube resembled that of phoronids, sessile animals that feed by lophophores and are regarded either very close relatives or a sub-group of brachiopods. *Paterimitra*, another mostly assembled fossil found in 2008 and described in 2009, had two symmetrical plates at the bottom, like brachiopod valves but not fully enclosing the animal's body.

At their peak in the Paleozoic, the brachiopods were among the most abundant filter-feeders and reef-builders, and occupied other ecological niches, including swimming in the jet-propulsion style of scallops. However, after the Permian–Triassic extinction event, informally known as the "Great

Dying", brachiopods recovered only a third of their former diversity. It was often thought that brachiopods were actually declining in diversity, and that in some way bivalves out-competed them. However, in 1980, Gould and Calloway produced a statistical analysis that concluded that: both brachiopods and bivalves increased all the way from the Paleozoic to modern times, but bivalves increased faster; the Permian–Triassic extinction was moderately severe for bivalves but devastating for brachiopods, so that brachiopods for the first time were less diverse than bivalves and their diversity after the Permian increased from a very low base; there is no evidence that bivalves out-competed brachiopods, and short-term increases or decreases for both groups appeared synchronously. In 2007 Knoll and Bambach concluded that brachiopods were one of several groups that were most vulnerable to the Permian–Triassic extinction, as all had calcareous hard parts (made of calcium carbonate) and had low metabolic rates and weak respiratory systems.

A dense assemblage of the Ordovician species *Cincinnetina meeki* (Miller, 1875)

Brachiopod fossils have been useful indicators of climate changes during the Paleozoic era. When global temperatures were low, as in much of the Ordovician, the large difference in temperature between equator and poles created different collections of fossils at different latitudes. On the other hand, warmer periods, such much of the Silurian, created smaller difference in temperatures, and all seas at the low to middle latitudes were colonized by the same few brachiopod species.

Evolutionary Family Tree

Productid brachiopod ventral valve; Roadian, Guadalupian (Middle Permian); Glass Mountains, Texas

Deuterostomes or Protostomes

From about the 1940s to the 1990s, family trees based on embryological and morphological features placed brachiopods among or as a sister group to the deuterostomes. a super-phylum that includes chordates and echinoderms. Closer examination has found difficulties in the grounds on which brachiopods were affiliated with deuterostomes:

- Radial cleavage in the earliest divisions of the egg appears to be the original condition for the ancestral bilaterians, in the earliest Ecdysozoa and possibly in the earliest Eutrochozoa, a major sub-group of the Lophotrochozoa. Hence radial cleavage does not imply that brachiopods are affiliated with deuterostomes.
- The traditional view is that the coelom(s) in deuterostomes and protostomes form by different process, called enterocoely and schizocoely respectively. However, research since the early 1990s has found significant exceptions. Both types of coelom construction appear among brachiopods, and therefore do not imply that brachiopods are deuterostomes.
- The terms "deuterostomes" and "protostomes" originally defined distinct ways of forming the mouth from the blastopore, a depression that appears in an early stage of the embryo. However, some "protostomes" forming the mouth using a process more like that typical of deuterostomes. Hence forming the mouth via a deuterostome-like process does not imply that brachiopods are affiliated with deuterostomes.

Nielsen views the brachiopods and closely related phoronids as affiliated with the deuterostome pterobranchs because their lophophores are driven by one cilium per cell, while those of bryozoans, which he regards as protosomes, have multiple cilia per cell. However, pterobranchs are hemichordates and probably closely related to echinoderms, and there is no evidence that the latest common ancestor of pterobranchs and other hemichordates or the latest common ancestor of hemichordates and echinoderms was sessile and fed by means of tentacles.

From 1988 onwards analyses based on molecular phylogeny, which compares biochemical features such as similarities in DNA, have placed brachiopods among the Lophotrochozoa, a protostome super-phylum that includes molluscs, annelids and flatworms but excludes the other protostome super-phylum Ecdysozoa, whose members include arthropods. This conclusion is unanimous among molecular phylogeny studies, which use a wide selection of genes: rDNA, Hox genes, mitochondrial protein genes, single nuclear protein genes and sets of nuclear protein genes.

Some combined studies in 2000 and 2001, using both molecular and morphological data, support brachiopods as Lophotrochozoa, while others in 1998 and 2004 concluded that brachiopods were deuterostomes.

Relationship with Other Lophotrochozoans

The phoronids feed with a lophophore, burrow or encrust on surfaces, and build three-layered tubes made of polysaccharide, possibly chitin, mixed with particles with seabed material. Traditionally they have been regarded as a separate phylum, but increasingly detailed molecular phylogeny studies between 1997 and 2000 have concluded that phoronids are a sub-group of brachiopods. However, an analysis in 2005 concluded that phoronids are a sub-group of bryozoans.

While all molecular phylogeny studies and half the combined studies until 2008 conclude that brachiopods are lophotrochozoans, they could not identify which lophotrochozoan phylum were the closest relatives of brachiopods—except phoronids, which are a sub-group of brachiopods. However, in 2008 two analyses found that brachiopods' closest lophotrochozoan relatives were nemertines. The authors found this surprising, since nemertines have spiral cleavage in the early stages of cell division and form a trochophore larva, while brachiopods have radial cleavage and a larva that shows no sign of having evolved from a trochophore. Another study in 2008 also concluded that brachiopods are closely related to nemertines, casting doubt on the idea that brachiopods are part of a clade Lophophorata of lophophore-feeding animals within the lophotrochozoans.

Bryozoa

Bryozoa (also known as the Polyzoa, Ectoprocta or commonly as moss animals), are a phylum of aquatic invertebrate animals. Typically about 0.5 millimetres (0.020 in) long, they are filter feeders that sieve food particles out of the water using a retractable lophophore, a "crown" of tentacles lined with cilia. Most marine species live in tropical waters, but a few occur in oceanic trenches, and others are found in polar waters. One class lives only in a variety of freshwater environments, and a few members of a mostly marine class prefer brackish water. Over 4,000 living species are known. One genus is solitary and the rest are colonial.

The phylum was originally called "Polyzoa", but this term was superseded by "Bryozoa" in 1831. Another group of animals discovered subsequently, whose filtering mechanism looked similar, was also included in "Bryozoa" until 1869, when the two groups were noted to be very different internally. The more recently discovered group was given the name Entoprocta, while the original "Bryozoa" were called "Ectoprocta". However, "Bryozoa" has remained the more widely used term for the latter group.

Individuals in bryozoan (ectoproct) colonies are called zooids, since they are not fully independent animals. All colonies contain autozooids, which are responsible for feeding and excretion. Colonies of some classes have various types of non-feeding specialist zooids, some of which are hatcheries for fertilized eggs, and some classes also have special zooids for defense of the colony. The class Cheilostomata have the largest number of species, possibly because they have the widest range of specialist zooids. A few species can creep very slowly by using spiny defensive zooids as legs. Autozooids supply nutrients to non-feeding zooids by channels that vary between classes. All zooids, including those of the solitary species, consist of a *cystid* that provides the body wall and produces the exoskeleton and a *polypide* that contains the internal organs and the lophophore or other specialist extensions. Zooids have no special excretory organs, and the polypides of autozooids are scrapped when the polypides become overloaded by waste products; usually the body wall then grows a replacement polypide. In autozooids the gut is U-shaped, with the mouth inside the "crown" of tentacles and the anus outside it. Colonies take a variety of forms, including fans, bushes and sheets. The Cheilostomata produce mineralized exoskeletons and form single-layered sheets which encrust over surfaces.

Zooids of all the freshwater species are simultaneous hermaphrodites. Although those of many marine species function first as males and then as females, their colonies always contain a combination of zooids that are in their male and female stages. All species emit sperm into the water. Some also release ova into the water, while others capture sperm via their tentacles to fertilize their ova internally. In some species the larvae have large yolks, go to feed, and quickly settle on a surface. Others produce larvae that have little yolk but swim and feed for a few days before settling. After settling, all larvae undergo a radical metamorphosis that destroys and rebuilds almost all the internal tissues. Freshwater species also produce statoblasts that lie dormant until conditions are favorable, which enables a colony's lineage to survive even if severe conditions kill the mother colony.

Predators of marine bryozoans include nudibranchs (sea slugs), fish, sea urchins, pycnogonids, crustaceans, mites and starfish. Freshwater bryozoans are preyed on by snails, insects, and fish. In Thailand, many populations of one freshwater species have been wiped out by an introduced species of snail. A fast-growing invasive bryozoan off the northeast and northwest coasts of the USA has reduced kelp forests so much that it has affected local fish and invertebrate populations. Bryozoans have spread diseases to fish farms and fishermen. Chemicals extracted from a marine bryozoan species have been investigated for treatment of cancer and Alzheimer's disease, but analyses have not been encouraging.

Mineralized skeletons of bryozoans first appear in rocks from Early Ordovician period, making it the last major phylum to appear in the fossil record. This has led researchers to suspect that bryozoans had arisen earlier but were initially unmineralized, and may have differed significantly from fossilized and modern forms. Early fossils are mainly of erect forms, but encrusting forms gradually became dominant. It is uncertain whether the phylum is monophyletic. Bryozoans' evolutionary relationships to other phyla are also unclear, partly because scientists' view of the family tree of animals is mainly influenced by better-known phyla. Both morphological and molecular phylogeny analyses disagree over bryozoans' relationships with entoprocts, about whether bryozoans should be grouped with brachiopods and phoronids in Lophophorata, and whether bryozoans should be considered protostomes or deuterostomes.

Description

Distinguishing Features

Bryozoans, phoronids and brachiopods strain food out of the water by means of a lophophore, a "crown" of hollow tentacles. Bryozoans form colonies consisting of clones called zooids that are typically about 0.5 millimetres (0.020 in) long. Phoronids resemble bryozoan zooids but are 2 to 20 centimetres (0.79 to 7.87 in) long and, although they often grow in clumps, do not form colonies consisting of clones. Brachiopods, generally thought to be closely related to bryozoans and phoronids, are distinguished by having shells rather like those of bivalves. All three of these phyla have a coelom, an internal cavity lined by mesothelium. Some encrusting bryozoan colonies with mineralized exoskeletons look very like small corals. However, bryozoan colonies are founded by an ancestrula, which is round rather than shaped like a normal zooid of that species. On the other hand, the founding polyp of a coral has a shape like that of its daughter polyps, and coral zooids have no coelom or lophophore.

Entoprocts, another phylum of filter-feeders, look rather like bryozoans but their lophophore-like feeding structure has solid tentacles, their anus lies inside rather than outside the base of the "crown" and they have no coelom.

Types of Zooid

All bryozoans are colonial except for one genus, *Monobryozoon*. Individual members of a bryozoan colony are about 0.5 millimetres (0.020 in) long and are known as *zooids*, since they are not fully independent animals. All colonies contain feeding zooids, known as autozooids, and those of some groups also contain non-feeding specialist heterozooids; colony members are genetically identical and co-operate, rather like the organs of larger animals. What type of zooid grows where in a colony is determined by chemical signals from the colony as a whole or sometimes in response to the scent of predators or rival colonies.

The bodies of all types have two main parts. The *cystid* consists of the body wall and whatever type of exoskeleton is secreted by the epidermis. The exoskeleton may be organic (chitin, polysaccharide or protein) or made of the mineral calcium carbonate. The body wall consists of the epidermis, basal lamina (a mat of non-cellular material), connective tissue, muscles, and the mesothelium which lines the coelom (main body cavity) – except that in one class, the mesothelium is split into two separate layers, the inner one forming a membranous sac that floats freely and contains the coelom, and the outer one attached to the body wall and enclosing the membranous sac in a pseudocoelom. The other main part of the bryozoan body, known as the *polypide* and situated almost entirely within the cystid, contains the nervous system, digestive system, some specialized muscles and the feeding apparatus or other specialized organs that take the place of the feeding apparatus.

The Bryozoa, known as "moss animals", currently contain about 5,000 known species. The different species can come together to form a colony that can attach to different kind of rock surfaces as well as shells and algae. Of the 5,000 species the 125 that have the most abundant population are known to attach themselves to the bottom of boats, pilings, piers and docks. These species are thought to be nuisance to boaters because they slow down the ship if they are attached to the bottom of the hull. However, there are species which have chemical compounds that can be used for medicine. One common species of the bryozoan is used for a serious anti cancer drug that is currently being used for testing.

Feeding Zooids

The most common type of zooid is the feeding autozooid, in which the polypide bears a "crown" of hollow tentacles called a lophophore, which captures food particles from the water. In all colonies a large percentage of zooids are autozooids, and some consist entirely of autozooids, some of which also engage in reproduction.

The basic shape of the "crown" is a full circle. In the class Phylactolaemata the crown appears U-shaped, but this impression is created by a deep dent in the rim of the crown, which has no gap in the fringe of tentacles. The sides of the tentacles bear fine hairs called cilia, whose beating drives a water current from the tips of the tentacles to their bases, where it exits. Food particles that collide with the tentacles are trapped by mucus, and further cilia on the inner surfaces of the tentacles convey the particles towards the mouth, which lies in the center of the base of the "crown". The method used by ectoprocts is known as "upstream collecting", as food particles are captured before they pass through the field of cilia that creates the feeding current. This method is also used by phoronids, brachiopods and pterobranchs.

The lophophore and mouth are mounted on a flexible tube, called the "invert" because it can be turned inside-out and withdrawn into the polypide, rather like the finger of a rubber glove; in this position the lophophore lies inside the invert and is folded like the spokes of an umbrella. The invert is withdrawn, sometimes within 60 milliseconds, by a pair of retractor muscles that are anchored at the far end of the cystid. Sensors at the tips of the tentacles may check for signs of danger before the invert and lophophore are fully extended. Extension is driven by an increase in internal fluid pressure, which species with flexible exoskeletons produce by contracting circular muscles that lie just inside the body wall, while species with a membranous sac use circular muscles to squeeze this. Some species with rigid exoskeletons have a flexible membrane that replaces part of the exoskeleton, and transverse muscles anchored on the far side of the exoskeleton increase the fluid pressure by pulling the membrane inwards. In others there is no gap in the protective skeleton, and the transverse muscles pull on a flexible sac which is connected to the water outside by a small pore; the expansion of the sac increases the pressure inside the body and pushes the invert and lophophore out. In some species the retracted invert and lophophore are protected by an operculum ("lid"), which is closed by muscles and opened by fluid pressure. In one class, a hollow lobe called the "epistome" overhands the mouth.

The gut is U-shaped, running from the mouth, in the center of the lophophore, down into the animal's interior and then back to the anus, which is located on the invert, outside and usually below the lophophore. A network of strands of mesothelium called "funiculi" ("little ropes") connects the mesothelium covering the gut with that lining the body wall. The wall of each strand is made of mesothelium, and surrounds a space filled with fluid, thought to be blood. A colony's zooids are connected, enabling autozooids to share food with each other and with any non-feeding heterozooids. The method of connection varies between the different classes of bryozoans, ranging from quite large gaps in the body walls to small pores through which nutrients are passed by funiculi.

There is a nerve ring round the pharynx (throat) and a ganglion that serves as a brain to one side of this. Nerves run from the ring and ganglion to the tentacles and to the rest of the body. Bryozoans have no specialized sense organs, but cilia on the tentacles act as sensors. Members of the genus *Bugula* grow towards the sun, and therefore must be able to detect light. In colonies of some species, signals are transmitted between zooids through nerves that pass through pores in the body walls, and coordinate activities such as feeding and the retraction of lophophores.

The solitary individuals of *Monobryozoon* are autozooids with pear-shaped bodies. The wider ends have up to 15 short, muscular projections by which the animals anchor themselves to sand or gravel and pull themselves through the sediments.

Avicularia and Vibracula

Some authorities use the term *avicularia* (pl. of *avicularium*) to refer to any type of zooid in which the lophophore is replaced by an extension that serves some protective function, while others restrict the term to those that defend the colony by snapping at invaders and small predators, killing some and biting the appendages of others. In some species the snapping zooids are mounted on a peduncle (stalk), their bird-like appearance responsible for the term – Charles Darwin described these as like "the head and beak of a vulture in miniature, seated on a neck and capable of movement". Stalked avicularia are placed upside-down on their stalks. The "lower jaws" are modified versions of the opercula that protect the retracted lophophores in autozooids of some species, and are snapped shut "like a mousetrap" by similar muscles, while the beak-shaped upper jaw is the inverted body wall. In other species the avicularia are stationary box-like zooids laid the normal way up, so that the modified operculum snaps down against the body wall. In both types the modified operculum is opened by other muscles that attach to it, or by internal muscles that raise the fluid pressure by pulling on a flexible membrane. The actions of these snapping zooids are controlled by small, highly modified polypides that are located inside the "mouth" and bear tufts of short sensory cilia. These zooids appear in various positions: some take the place of autozooids, some fit into small gaps between autozooids, and small avicularia may occur on the surfaces of other zooids.

In vibracula, regarded by some as a type of avicularia, the operculum is modified to form a long bristle that has a wide range of motion. They may function as defenses against predators and invaders, or as cleaners. In some species that form mobile colonies, vibracula around the edges are used as legs for burrowing and walking.

Other Types of Colonial Zooid

Kenozooids (from Greek meaning "empty") consist only of the body wall and funicular strands crossing the interior, and no polypide. In some species they form the stems of branching structures, while in others they act as spacers that enable colonies to grow quickly in a new direction.

Spinozooids form defensive spines, and sometimes appear on top of autozooids. Gonozooids act as brood chambers for fertilized eggs. Some species have miniature nanozooids with small single-tentacled polypides, and these may grow on other zooids or within the body walls of autozooids that have degenerated.

Colony Forms and Composition

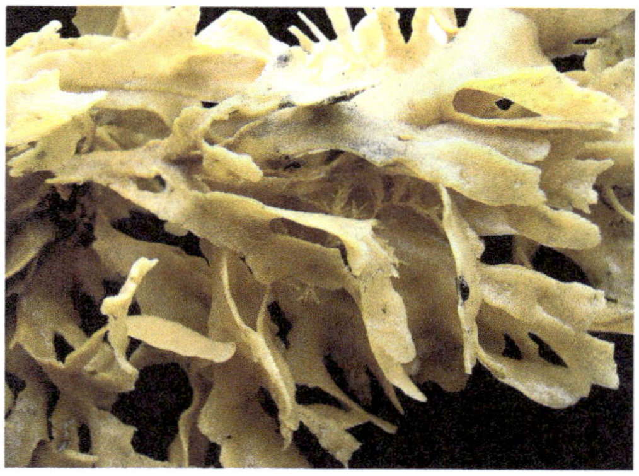

A colony of the modern marine bryozoan *Flustra foliacea*.

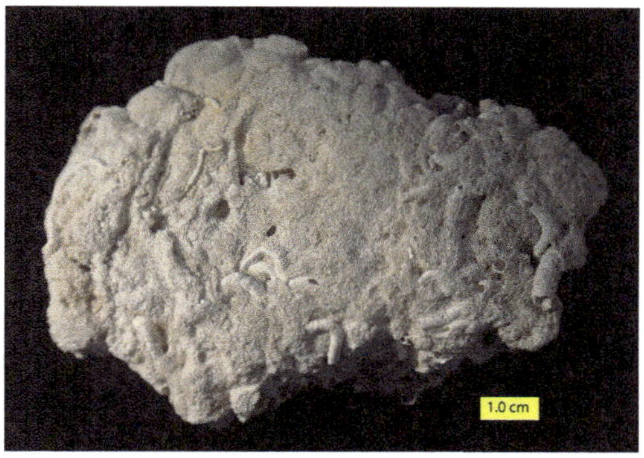

Cheilostome bryozoan with serpulid tubes; Recent; Cape Cod Bay, Duck Creek, near Wellfleet, Massachusetts.

Although zooids are microscopic, colonies range in size from 1 centimeter (0.39 in) to over 1 meter (3.3 ft). However, the majority are under 10 centimeters (3.9 in) across. The shapes of colonies vary widely, depend on the pattern of budding by which they grow, the variety of zooids present and the type and amount of skeletal material they secrete.

Some marine species are bush-like or fan-like, supported by "trunks" and "branches" formed by kenozooids, with feeding autozooids growing from these. Colonies of these types are generally unmineralized but may have exoskeletons made of chitin. Others look like small corals, producing heavy lime skeletons. Many species form colonies which consist of sheets of autozooids. These sheets may form leaves, tufts or, in the genus *Thalmoporella*, structures that resemble an open head of lettuce.

The most common marine form, however, is encrusting, in which a one-layer sheet of zooids spreads over a hard surface or over seaweed. Some encrusting colonies may grow to over 50 centimeters (1.6 ft) and contain about 2,000,000 zooids. These species generally have exoskeletons reinforced with calcium carbonate, and the openings through which the lophophores protrude are on the top or outer surface. The moss-like appearance of encrusting colonies is responsible for the phylum's name (Ancient Greek words *bryos* meaning "moss" and *zoon* meaning "animal"). Large colonies of encrusting species often have "chimneys", gaps in the canopy of lophophores, through which they swiftly expel water that has been sieved, and thus avoid re-filtering water that is already exhausted. They are formed by patches of non-feeding heterozooids. New chimneys appear near the edges of expanding colonies, at points where the speed of the outflow is already high, and do not change position if the water flow changes.

Some freshwater species secrete a mass of gelatinous material, up to 1 meter (3.3 ft) in diameter, to which the zooids stick. Other freshwater species have plant-like shapes with "trunks" and "branches", which may stand erect or spread over the surface. A few species can creep at about 2 centimeters (0.79 in) per day.

Each colony grows by asexual budding from a single zooid known as the ancestrula, which is round rather than shaped like a normal zooid. This occurs at the tips of "trunks" or "branches" in forms that have this structure. Encrusting colonies grow round their edges. In species with calcareous exoskeletons, these do not mineralize until the zooids are fully grown. Colony lifespans range from one to about 12 years, and the short-lived species pass though several generations in one season.

Species that produce defensive zooids do so only when threats have already appeared, and may do so within 48 hours. The theory of "induced defenses" suggests that production of defenses is expensive and that colonies which defend themselves too early or too heavily will have reduced growth rates and lifespans. This "last minute" approach to defense is feasible because the loss of zooids to a single attack is unlikely to be significant. Colonies of some encrusting species also produce special heterozooids to limit the expansion of other encrusting organisms, especially other bryozoans. In some cases this response is more belligerent if the opposition is smaller, which suggests that zooids on the edge of a colony can somehow sense the size of the opponent. Some species consistently prevail against certain others, but most turf wars are indecisive and the combatants soon turn to growing in uncontested areas. Bryozoans competing for territory do not use the sophisticated techniques employed by sponges or corals, possibly because the shortness of bryozoan lifespans makes heavy investment in turf wars unprofitable.

Bryozoans have contributed to carbonate sedimentation in marine life since the Ordovician period. Bryozoans take responsibility for many of the colony forms, which have evolved in different taxonomic groups and vary in sediment producing ability. The nine basic bryozoan colony-forms include: encrusting, dome-shaped, palmate, foliose, fenestrate, robust branching, delicate branching, articulated and free-living. Most of these sediments come from two distinct groups of colonies: domal, delicate branching, robust branching and palmate; and fenestrate. Fenestrate colonies generate rough particles both as sediment and components of stomatoporiods coral reefs. The delicate colonies however, create both coarse sediment and form the cores of deep-water, subphotic biogenic mounds. Nearly all post- bryozoan sediments are made up of growth forms, with the addition to free-living colonies which include significant numbers of various colonies. "In contrast to the Palaeozoic, post-Palaeozoic bryozoans generated sediment varying more widely with the size of their grains; they grow as they moved from mud, to sand, to gravel."

Taxonomy

Peronopora, a trepostome bryozoan from the Whitewater Formation (Upper Ordovician) of eastern Indiana.

Evactinopora bryozoan found in Jefferson County, Missouri, United States; from the permanent collection of The Children's Museum of Indianapolis.

The phylum was originally called "Polyzoa", but this name was soon replaced by Ehrenberg's term "Bryozoa". The name "Bryozoa" was originally applied only to the animals also known as "Ectoprocta", in which the anus lies outside the "crown" of tentacles. After the discovery of the Entoprocta, in which the anus lies within a "crown" of tentacles, the name "Bryozoa" was used at phylum level to include the two classes Ectoprocta and Entoprocta. However, in 1869 Hinrich Nitsche regarded the two groups as quite distinct for a variety of reasons, and coined the name "Ectoprocta" for Ehrenberg's "Bryozoa". Despite their apparently similar methods of feeding, they differed markedly anatomically; in addition to the different positions of the anus, ectoprocts have hollow tentacles and a coelom, while entoprocts have solid tentacles and no coelom. Hence the two groups are now widely regarded as separate phyla, and the name "Bryozoa" is now synonymous with "Ectoprocta". This has remained the majority view ever since, although most publications have preferred the name "Bryozoa" rather than "Ectoprocta". Nevertheless, some notable scientists have continued to regard the "Ectoprocta" and Entoprocta as close relatives and group them under "Bryozoa".

The ambiguity about the scope of the name "Bryozoa" led to proposals in the 1960s and 1970s that it should be avoided and the unambiguous term "Ectoprocta" should be used. However, the change would have made it harder to find older works about in which the phylum was called "Bryozoa", and the desire to avoid ambiguity, if applied consistently to all classifications, would have necessitated renaming of several other phyla and many lower-level groups. In practice, zoological naming of split or merged groups of animals is complex and not completely consistent. Works since 2000 have used various names to resolve the ambiguity, including: "Bryozoa", "Ectoprocta", "Bryozoa (Ectoprocta)", and "Ectoprocta (Bryozoa)". Some have used more than one approach in the same work.

The common name "moss animals" is based on the Greek (moss) and (animals), and refers to the mossy appearance of encrusting species.

Up until recently (2008) there were "inadequately known and misunderstood type species belonging to the Cyclostome Bryozoan family Oncousoeciidae." (Taylor, Zaton 2008) Modern research and experiments have been done using low-vacuum scanning electron microscopy of uncoated type material to critically examine and perhaps revise the taxonomy of three genera belonging to this family, including *Oncousoecia*, *Microeciella*, and *Eurystrotos*. This method permits data to be obtained that would be difficult to recognize with an optical microscope. The valid type species of *Oncousoecia* was found to be *Oncousoecia lobulata*. This interpretation stabilizes *Oncousoecia* by establishing a type species that corresponds to the general usage of the genus. Fellow Oncousoeciid *Eurystrotos* is now believed to be not conspecific with *O. lobulata*, as previously suggested, but shows enough similarities to be considered a junior synonym of *Oncousoecia*. *Microeciella suborbicularus* has also been recently distinguished from *O. lobulata* and *O. dilatans*, using this modern method of low vacuum scanning, with which it has been inaccurately synonymized with in the past. A new genus has also been recently discovered called *Junerossia* in the family Stomachetosellidae, along with 10 relatively new species of bryozoa such as *Alderina flaventa*, *Corbulella extenuata*, *Puellina septemcryptica*, *Junerossia copiosa*, *Calyptotheca kapaaensis*, *Bryopesanser serratus*, *Cribellopora souleorum*, *Metacleidochasma verrucosa*, *Disporella compta*, and *Favosipora adunca*.

Fossil Record

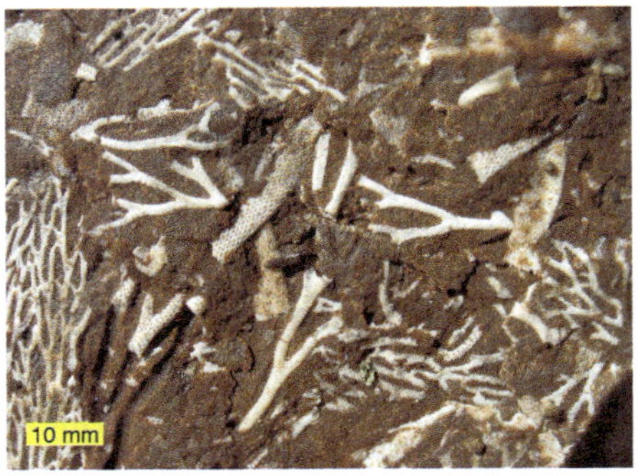

Bryozoan fossils in an Upper Ordovician oil shale (kukersite), northern Estonia.

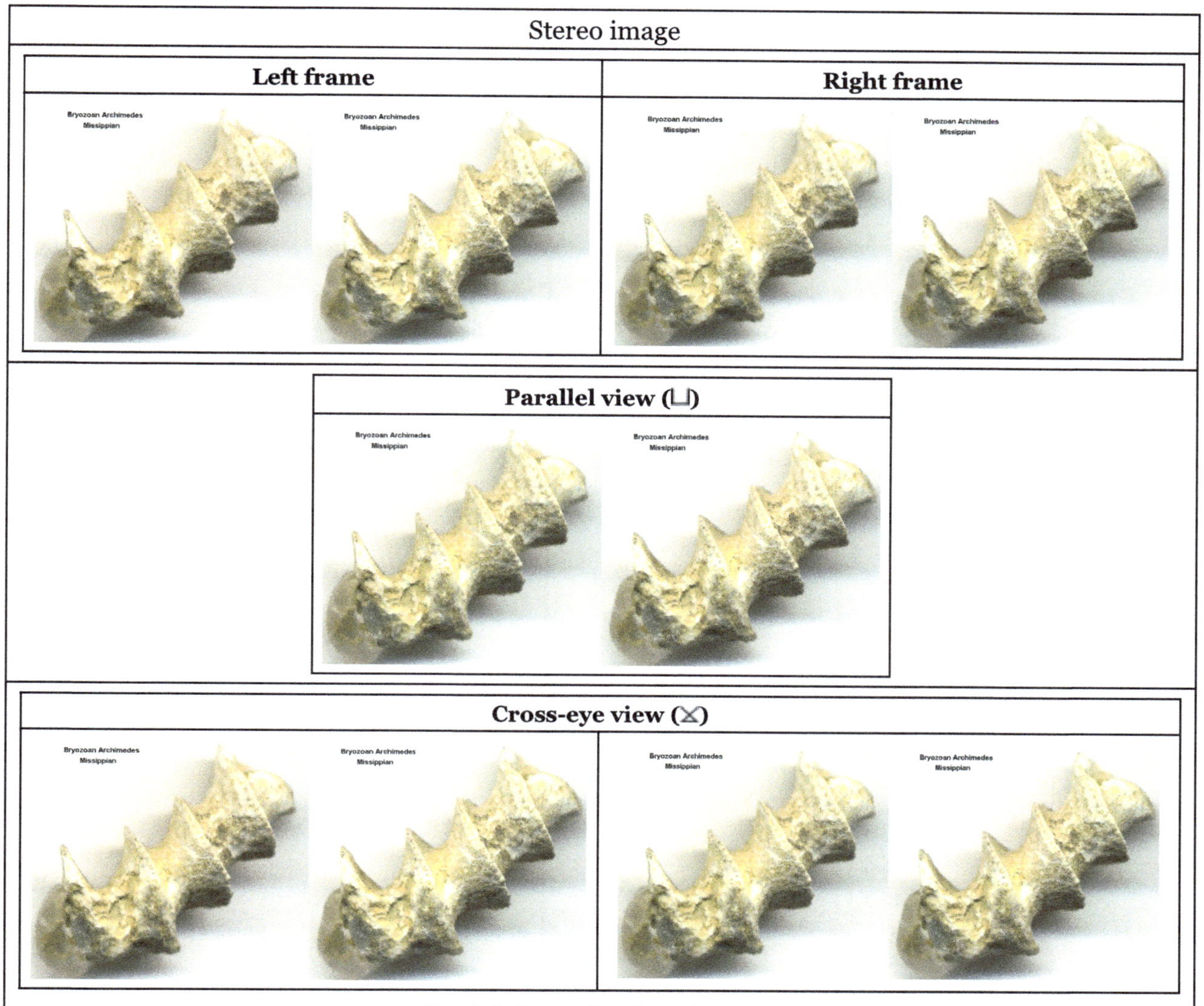

Fossilized skeleton of Archimedes Bryozoan

Fossils of about 15,000 bryozoan species have been found. Bryozoans are among the three dominant groups of Paleozoic fossils. The oldest species with a mineralized skeleton occurs in the Lower Ordovician. It is likely that the first bryozoans appeared much earlier and were entirely soft-bodied, and the Ordovician fossils record the appearance of mineralized skeletons in this phylum. By the Arenigian stage of the Early Ordovician period, about 480 million years ago, all the modern orders of stenolaemates were present, and the ctenostome order of gymnolaemates had appeared by the Middle Ordovician, about 465 million years ago. The Early Ordovician fossils may also represent forms that had already become significantly different from the original members of the phylum. Ctenostomes with phosphatized soft tissue are known from the Devonian. Other types of filter feeders appeared around the same time, which suggests that some change made the environment more favorable for this lifestyle. Fossils of cheilostomates, another order of gymnolaemates, first appear in the Mid Jurassic, about 172 million years ago, and these have been the most abundant and diverse bryozoans from the Cretaceous to the present. Evidence compiled from the last 100 million years show that cheilostomates consistently grew over cyclostomates in territorial struggles, which may help to explain how cheilostomates replaced cyclostomates as the dominant marine bryozoans. Marine fossils from the Paleozoic era, which ended 251 million years ago, are mainly of erect forms, those from the Mesozoic are fairly equally divided by erect and encrusting forms, and more recent ones are predominantly encrusting. Fossils of the soft, freshwater phylactolaemates are very rare, appear in and after the Late Permian (which began about 260 million years ago) and consist entirely of their durable statoblasts. There are no known fossils of freshwater members of other classes.

Evolutionary Family Tree

Scientists are divided about whether the Bryozoa (Ectoprocta) are a monophyletic group (whether they include all and only a single ancestor species and all its descendants), about what are the phylum's closest relatives in the family tree of animals, and even about whether they should be regarded as members of the protostomes or deuterostomes, the two major groups that account for all moderately complex animals.

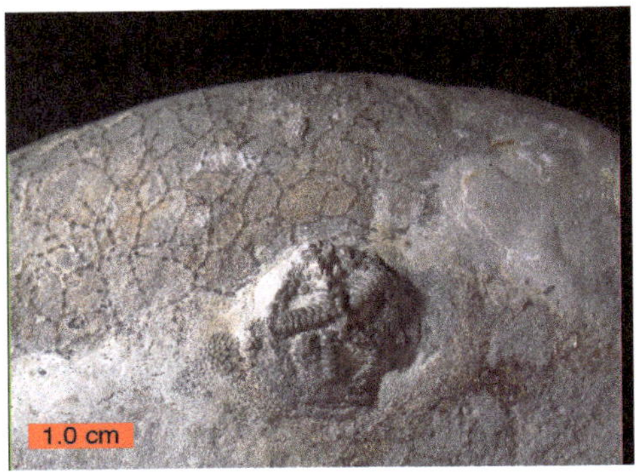

An Upper Ordovician cobble with the edrioasteroid *Cystaster stellatus* and the thin branching cyclostome bryozoan *Corynotrypa*. Kope Formation, northern Kentucky.

Ropalonaria venosa, an etching trace fossil of a Late Ordovician ctenostome bryozoan on a strophomenid brachiopod valve; Cincinnatian of southeastern Indiana.

The traditional view is that the Bryozoa are a monophyletic group, in which the class Phylactolaemata is most closely related to Stenolaemata and Ctenostomata, the classes that appear earliest in the fossil record. However, in 2005 a molecular phylogeny study that focused on phylactolaemates concluded that these are more closely related to the phylum Phoronida, and especially to the only phoronid species that is colonial, than they are to the other ectoproct classes. That implies that the Entoprocta are not monophyletic, as the Phoronida are a subgroup of ectoprocts but the standard definition of Entoprocta excludes the Phoronida. In 2009 another molecular phylogeny study, using a combination of genes from mitochondria and the cell nucleus, concluded that Bryozoa is a monophyletic phylum, in other words includes all the descendants of a common ancestor that is itself a bryozoan. The analysis also concluded that the classes Phylactolaemata, Stenolaemata and Gymnolaemata are also monophyletic, but could not determine whether Stenolaemata are more closely related to Phylactolaemata or Gymnolaemata. The Gymnolaemata are traditionally divided into the soft-bodied Ctenostomata and mineralized Cheilostomata, but the 2009 analysis considered it more likely that neither of these orders is monophyletic and that mineralized skeletons probably evolved more than once within the early Gymnolaemata.

Phaenopora superba, a ptilodictyine bryozoan from the Silurian of Ohio.

Bryozoans' relationships with other phyla are uncertain and controversial. Traditional phylogeny, based on anatomy and on the development of the adult forms from embryos, has produced no enduring consensus about the position of ectoprocts. Attempts to reconstruct the family tree of animals have largely ignored ectoprocts and other "minor phyla", which have received little scientific study because they are generally tiny, have relatively simple body plans, and have little impact on human economies – despite the fact that the "minor phyla" include most of the variety in the evolutionary history of animals. In the opinion of Ruth Dewel, Judith Winston and Frank McKinney, "Our standard interpretation of bryozoan morphology and embryology is a construct resulting from over 100 years of attempts to synthesize a single framework for all invertebrates," and takes little account of some peculiar features of ectoprocts. In ectoprocts, all of the larva's internal organs are destroyed during the metamorphosis to the adult form and the adult's organs are built from the larva's epidermis and mesoderm, while in other bilaterians some organs including the gut are built from endoderm. In most bilaterian embryos the blastopore, a dent in the outer wall, deepens to become the larva's gut, but in ectoprocts the blastopore disappears and a new dent becomes the point from which the gut grows. The ectoproct coelom is formed by neither of the processes used by other bilaterians, enterocoely, in which pouches that form on the wall of the gut become separate cavities, nor schizocoely, in which the tissue between the gut and the body wall splits, forming paired cavities.

Molecular phylogeny, which attempts to work out the evolutionary family tree of organisms by comparing their biochemistry and especially their genes, has done much to clarify the relationships between the better-known invertebrate phyla. However, the shortage of genetic data about "minor phyla" such as bryozoans and entoprocts has left their relationships to other groups unclear.

When entoprocts were discovered in the 19th century, they and bryozoans (ectoprocts) were regarded as classes within the phylum Bryozoa, because both groups were sessile animals that filter-fed by means of a crown of tentacles that bore cilia. However, from 1869 onwards increasing awareness of differences, including the position of the entoproct anus inside the feeding structure and the difference in the early pattern of division of cells in their embryos, caused scientists to regard the two groups as separate phyla, and "Bryozoa" became just an alternative name for ectoprocts, in which the anus is outside the feeding organ. A series of molecular phylogeny studies from 1996 to 2006 have also concluded that bryozoans (ectoprocts) and entoprocts are not sister groups. However, two well-known zoologists, Claus Nielsen and Thomas Cavalier-Smith, maintain on anatomical and developmental grounds that bryozoans and entoprocts are member of the same phylum, Bryozoa. A molecular phylogeny study in 2007 also supported this old idea, while its conclusions about other phyla agreed with those of several other analyses.

By 1891 bryozoans (ectoprocts) were grouped with phoronids in a super-phylum called "Tentaculata". In the 1970s comparisons between phoronid larvae and the cyphonautes larva of some gymnolaete bryozoans produced suggestions that the bryozoans, most of which are colonial, evolved from a semi-colonial species of phoronid. Brachiopods were also assigned to the "Tentaculata", which were renamed Lophophorata as they all use a lophophore for filter feeding. Although the majority of scientists accept this, Claus Nielsen thinks these similarities are superficial. The Lophophorata are usually defined as animals with a lophophore, a three-part coelom and a U-shaped gut. In Nielsen's opinion, phoronids' and brachiopods' lophophores are more like those of pterobranchs, which are members of the phylum Hemichordata. Bryozoan's tentacles bear cells with multiple cilia, while the corresponding cells of phoronids', brachiopods' and pterobranchs' lophophores have one cilium per cell; and bryozoan tentacles have no hemal canal ("blood vessel"), which those of the other three phyla have.

If the grouping of bryozoans with phoronids and brachiopods into Lophophorata is correct, the next issue is whether the Lophophorata are protostomes, along with most invertebrate phyla, or deuterostomes, along with chordates, hemichordates and echinoderms. The traditional view was that lophophorates were a mix of protostome and deuterostome features. Research from the 1970s onwards suggested they were deuterostomes, because of some features that were thought characteristic of deuterostomes: a three-part coelom; radial rather than spiral cleavage in the development of the embryo; and formation of the coelom by enterocoely. However the coelom of ectoproct larvae shows no sign of division into three sections, and that of adult ectoprocts is different from that of other coelomate phyla as it is built anew from epidermis and mesoderm after metamorphosis has destroyed the larval coelom.

Molecular phylogeny analyses from 1995 onwards, using a variety of biochemical evidence and analytical techniques, placed the lophophorates as protostomes and closely related to annelids and molluscs in a super-phylum called Lophotrochozoa. "Total evidence" analyses, which used both morphological features and a relatively small set of genes, came to various conclusions, mostly favoring a close relationship between lophophorates and Lophotrochozoa. A study in 2008, using a larger set of genes, concluded that the lophophorates were closer to the Lophotrochozoa than to deuterostomes, but also that the lophophorates were not monophyletic. Instead, it concluded that brachiopods and phoronids formed a monophyletic group, but bryozoans (ectoprocts) were closest to entoprocts, supporting the original definition of "Bryozoa".

They are also the only major phylum of exclusively clonal animals and are all colonial. They are colonies of modular units known as zooids. Because they thrive in colonies, colonial growth allows them to develop unrestricted variations in form. Despite this, only a small number of basic growth forms have been found and have commonly reappeared throughout the history of the bryozoa.

The phylogenetic position of the ectoproct bryozoans remains uncertain, but it remains certain that they belong to the Protostomia and more specifically to the Lophoctrochozoa. This implies that the ectoproct larva is a trochophore with the corona being a homologue of the prototroch; this is supported from the similarity between the coronate larvae and the Type 1 pericalymma larvae of some molluscs and sipunculans, where the prototroch zone is expanded to cover the hyposphere.

A study of the mitochondrial DNA sequence suggests that the Bryozoa may be related to the Chaetognatha.

Physiology

Feeding and Excretion

Most species are filter feeders that sieve small particles, mainly phytoplankton (microscopic floating plants), out of the water. The freshwater species *Plumatella emarginata* feeds on diatoms, green algae, cyanobacteria, non-photosynthetic bacteria, dinoflagellates, rotifers, protozoa, small nematodes, and microscopic crustaceans. While the currents that bryozoans generate to draw food towards the mouth are well understood, the exact method of capture is still debated. All species also flick larger particles towards the mouth with a tentacle, and a few capture zooplankton (planktonic animals) by using their tentacles as cages. In addition the tentacles, whose surface area is increased by microvilli (small hairs and pleats), absorb organic compounds dissolved in the water. Unwanted particles may be flicked away by tentacles or shut out by closing the mouth. A study in 2008 showed that both encrusting and erect colonies fed more quickly and grew faster in gentle than in strong currents.

In some species the first part of the stomach forms a muscular gizzard lined with chitinous teeth that crush armored prey such as diatoms. Wave-like peristaltic contractions move the food through the stomach for digestion. The final section of the stomach is lined with cilia (minute hairs) that compress undigested solids, which then pass through the intestine and out through the anus.

There are no nephridia ("little kidneys") or other excretory organs in bryozoa, and it is thought that ammonia diffuses out through the body wall and lophophore. More complex waste products are not excreted but accumulate in the polypide, which degenerates after a few weeks. Some of the old polypide is recycled, but much of it remains as a large mass of dying cells containing accumulated wastes, and this is compressed into a "brown body". When the degeneration is complete, the cystid (outer part of the animal) produces a new polypide, and the brown body remains in the coelom, or in the stomach of the new polypide and is expelled next time the animal defecates.

Respiration and Circulation

There are no respiratory organs, heart or blood vessels. instead zooids absorb oxygen and eliminate carbon dioxide through diffusion. Bryozoa accomplish diffusion through the use of either a thin membrane (in the case of anascans and some polyzoa) or through psudopores located on the outer dermis of the zooid. The different bryozoan groups use various methods to share nutrients and oxygen between zooids: some have quite large gaps in the body walls, allowing the coelomic fluid to circulate freely; in others the funiculi (internal "little ropes") of adjacent zooids connect via small pores in the body wall.

Reproduction and Life Cycles

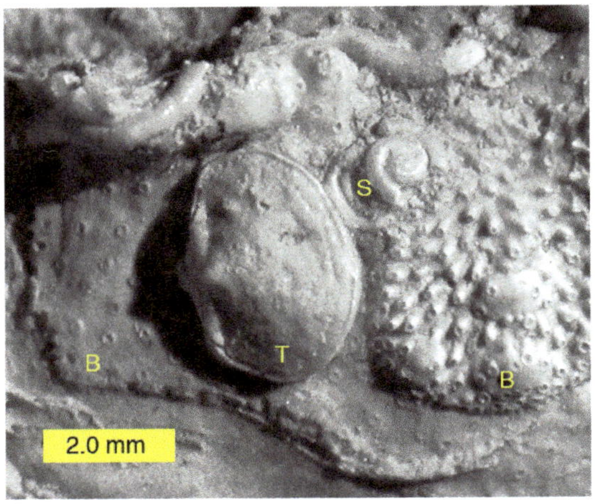

Encrusting cyclostome bryozoans (B), the one on the right showing swollen gonozooids;
T = thecideide brachiopod and S = sabellid worm tube; Jurassic of Poland.

Zooids of all phylactolaemate species are simultaneous hermaphrodites. Although those of many marine species are protandric, in other words function first as males and then as females, their colonies contain a combination of zooids that are in their male and female stages. In all species the ovaries develop on the inside of the body wall, and the testes on the funiculus connecting the stomach to the body wall. Eggs and sperm are released into the coelom, and sperm exit into the water through pores in the tips of some of the tentacles, and then are captured by the feeding currents of zooids that are producing eggs. Some species' eggs are fertilized externally after being released through a pore between two tentacles, which in some cases is at the tip of a small projection called the "intertentacular organ" in the base of a pair of tentacles. Others' are fertilized internally, in the intertentacular organ or in the coelom. In ctenostomes the mother provides a brood chamber for the fertilized eggs, and her polypide disintegrates, providing nourishment to the embryo. Stenolaemates produce specialized zooids to serve as brood chambers, and their eggs divide within this to produce up to 100 identical embryos.

The cleavage of bryozoan eggs is biradial, in other words the early stages are bilaterally symmetrical. It is unknown how the coleom forms, since the metamorphosis from larva to adult destroys all of the larva's internal tissues. In many animals the blastopore, an opening in the surface of the early embryo, tunnels through to form the gut. However, in bryozoans the blastopore closes, and a new opening develops to create the mouth.

Bryozoan larvae vary in form, but all have a band of cilia round the body which enables them to swim, a tuft of cilia at the top, and an adhesive sac that everts and anchors them when they settle on a surface. Some gymnolaemate species produce cyphonautes larvae which have little yolk but a well-developed mouth and gut, and live as plankton for a considerable time before settling. These larvae have triangular shells of chitin, with one corner at the top and the base open, forming a hood round the downward-facing mouth. In 2006 it was reported that the cilia of cyphonautes larvae use the same range of techniques as those of adults to capture food. Species that brood their embryos form larvae that are nourished by large yolks, have no gut and do not feed, and such larvae quickly settle on a surface. In all marine species the larvae produce cocoons in which they metamorphose completely after settling: the larva's epidermis becomes the lining of the coelom, and the internal tissues are converted to a food reserve that nourishes the developing zooid until it is ready to feed. The larvae of phylactolaemates produce multiple polypides, so that each new colony starts with several zooids. In all species the founder zooids then grow the new colonies by budding clones of themselves. In phylactolaemates, zooids die after producing several clones, so that living zooids are found only round the edges of a colony.

Phylactolaemates also reproduce asexually by a method that enables a colony's lineage to survive the variable and uncertain conditions of freshwater environments. Throughout summer and autumn they produce disc-shaped statoblasts, masses of cells that function as "survival pods" rather like the gemmules of sponges. Statoblasts form on the funiculus connected to the parent's gut, which nourishes them. As they grow, statoblasts develop protective bivalve-like shells made of chitin. When they mature, some statoblasts stick to the parent colony, some fall to the bottom ("sessoblasts"), some contain air spaces that enable them to float ("floatoblasts"), and some remain in the parent's cystid to re-build the colony if it dies. Statoblasts can remain dormant for considerable periods, and while dormant can survive harsh conditions such as freezing and desiccation. They can be transported across long distances by animals, floating vegetation, currents and winds, and even in the guts of larger animals. When conditions improve, the valves of the shell separate and the cells inside develop into a zooid that tries to form a new colony. *Plumatella emarginata* produces both "sessoblasts", which enable the lineage to control a good territory even if hard times decimate the parent colonies, and "floatoblasts", which spread to new sites. New colonies of *Plumatella repens* produce mainly "sessoblasts" while mature ones switch to "floatoblasts". A study estimated that one group of colonies in a patch measuring 1 square meter (11 sq ft) produced 800,000 statoblasts.

Cupuladriid Bryozoa are capable of both sexual and asexual reproduction. The sexually reproducing colonies (aclonal) are the result of a larval cupuladriid growing into an adult stage whereas the asexual colonies(clonal) are a result of a fragment of a colony of cupuladriids growing into its own colony. The different forms of reproduction in cupuladriids are achieved through a variety of methods depending on the morphology and classification of the zooid.

Ecology

Habitats and Distribution

Most marine species live in tropical waters at depths less than 100 meters (330 ft). However, a few have been found in deep-sea trenches, especially around cold seeps, and others near the poles. The great majority are sessile. Encrusting forms are much the commonest of these in shallow seas, but erect forms become more common as the depth increases. A few forms such as Cristatella can move, and an Antarctic species, Alcyonidium pelagosphaera, consists of floating colonies. The pelagic species is between 5.0 and 23.0 mm in diameter, has the shape of a hollow sphere and consists of a single layer of autozooids. It is still not known if these colonies are pelagic their whole life or only represents a temporarily and previously undescribed juvenile stage.

In 2014 it was reported that the bryozoan *Fenestrulina rugula* had become a dominant species in parts of Antarctica. Global warming has increased the rate of scouring by icebergs, and this species is particularly adept at recolonizing scoured areas.

The phylactolaemates live in all types of freshwater environment – lakes and ponds, rivers and streams, and estuaries – and are among the most abundant sessile freshwater animals. Some ctenostomes are exclusively freshwater while others prefer brackish water but can survive in freshwater. Scientists' knowledge of freshwater bryozoan populations in many parts of the world is incomplete, even in some parts of Europe. It was long thought that some freshwater species occurred worldwide, but since 2002 all of these have been split into more localized species.

Bryozoans are an immobile species typically residing on hard natural stone including, but not limited to: grains, shells, and rocks. Such sediment is customarily found in freshwater type marine niches, although a majority of Bryozoans develop in marine landscapes. It is not uncommon for colonies to grow on sediment and various other solid pseudo-rock formations. They are native to all five oceans making them a more cosmopolitan species.

Bryozoans are generally associated with the term colonies. Once a Bryozoa settles on a hard substance, after its larval phase, it is physically capable of reproducing asexually through budding. The term colony literally stems from the word clones. These colonies can grow thousands of individual zooids in a relatively short period of time. Even though colonies of zooids grow through asexual reproduction, Bryozoans are hermaphrodites and colonies are started through sexual reproduction. When colonies grow too large, however, they can split in two. This is the only case where asexual reproduction results in a new colony separate from its predecessor. Most colonies are stationary. Indeed, these colonies tend to be settled on immobile substances such as sediment and coarse substances. There are, in fact, other colonies, that predominantly reside in freshwater, that are able to move to a new locale; nevertheless, this movement is extremely slow, maximum 1 meter per hour, and demands energy.

Interactions with Non-human Organisms

lacelike *Membranipora membranacea*

Marine species are common on coral reefs, but seldom a significant proportion of the total biomass. In temperate waters, the skeletons of dead colonies form a significant component of shell gravels, and live ones are abundant in these areas. The marine lace-like bryozoan *Membranipora membranacea* produces spines in response to predation by several species of nudibranchs ("sea slugs"). Other predators on marine bryozoans include fish, sea urchins, pycnogonids, crustaceans, mites and starfish. In general marine echinoderms and molluscs eat masses of zooids by gouging pieces of colonies, breaking their mineralized "houses", while most arthropod predators on bryozoans eat individual zooids.

In freshwater, bryozoans are among the most important filter feeders, along with sponges and mussels. Freshwater bryozoans are attacked by many predators, including snails, insects, and fish.

In Thailand the introduced species *Pomacea canaliculata* (golden apple snail), which is generally a destructive herbivore, has wiped out phylactolaemate populations wherever it has appeared. *P. canaliculata* also preys on a common freshwater gymnolaemate, but with less devastating effect. Indigenous snails do not feed on bryozoans.

Several species of the hydroid family Zancleidae have symbiotic relationships with bryozoans, some of which are beneficial to the hydroids while others are parasitic. Modifications appear in the shapes of some these hydroids, for example smaller tentacles or encrustation of the roots by bryozoans. The bryozoan *Alcyonidium nodosum* protects the whelk *Burnupena papyracea* against predation by the powerful and voracious rock lobster *Jasus lalandii*. While whelk shells encrusted by the bryozoans are stronger than those without this reinforcement, chemical defenses produced by the bryozoans are probably the more significant deterrent.

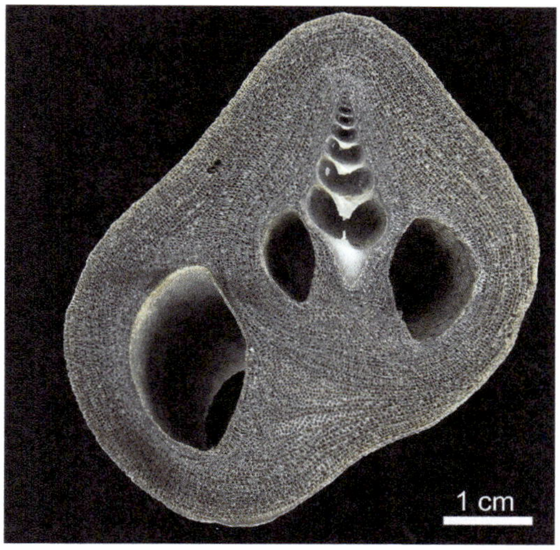

Mauritanian bryolith formed by circumrotatory growth of the bryozoan species *Acanthodesia commensale*

In the Banc d'Arguin offshore Mauritania the species *Acanthodesia commensale*, which is generally growing attached to gravel and hard-substrate, has formed a non-obligate symbiotic relationship with hermit crabs of the species *Pseudopagurus cf. granulimanus* resulting in egg-size structures known as bryoliths. Nucleating on an empty gastropod shell, the bryozoan colonies form multilamellar skeletal crusts that produce spherical encrustations and extend the living chamber of the hermit crab through helicospiral tubular growth.

Some phylactolaemate species are parasitized by a group of myxozoa that have also been found to cause Proliferative Kidney Disease, which is often fatal in salmonid fish, and has severely reduced wild fish populations in Europe and North America.

Membranipora membranacea, whose colonies feed and grow exceptionally fast in a wide range of current speeds, was first noticed in the Gulf of Maine in 1987 and quickly became the most abundant organism living on kelps. This invasion reduced the kelp population by breaking their fronds, so that its place as the dominant "vegetation" in some areas was taken by another invader, the large alga *Codium fragile tomentosoides*. These changes reduced the area of habitat available for local fish and invertebrates. *M. membranacea* has also invaded the northwest coast of the US A few freshwater species have been also found thousands of kilometers from their native ranges. Some may have been transported naturally as statoblasts. Others more probably were spread by humans, for example on imported water plants or as stowaways on ships.

Interaction with Humans

Fish farms and hatcheries have lost stock to Proliferative Kidney Disease, which is caused by one or more myxozoans that use bryozoans as alternate hosts.

Some fishermen in the North Sea have had to find other work because of a form of eczema (a skin disease) known as "Dogger Bank itch", caused by contact with bryozoans that have stuck to nets and lobster pots.

Marine bryozoans are often responsible for biofouling on ships' hulls, on docks and marinas, and on offshore structures. They are among the first colonizers of new or recently cleaned structures. Freshwater species are occasional nuisances in water pipes, drinking water purification equipment, sewage treatment facilities, and the cooling pipes of power stations.

A group of chemicals called bryostatins can be extracted from the marine bryozoan *Bugula neritina*. In 2001 pharmaceutical company GPC Biotech licensed Bryostatin 1 from Arizona State University for commercial development as a treatment for cancer. GPC Biotech canceled development in 2003, saying that Bryostatin 1 showed little effectiveness and some toxic side-effects. In January 2008 a clinical trial was submitted to the United States National Institutes of Health to measure the safety and effectiveness of Bryostatin 1 in the treatment of Alzheimer's Disease. However, no participants had been recruited by the end of December 2008, when the study was scheduled for completion. About 1,000 kilograms (2,200 lb) of bryozoans must be processed to extract 1 gram (0.035 oz) of bryostatin. As a result, synthetic equivalents have been developed that are simpler to produce and apparently at least as effective.

Chaetognatha

Chaetognatha, meaning *bristle-jaws*, and commonly known as arrow worms, is a phylum of predatory marine worms which are a major component of plankton worldwide. About 20% of the known species are benthic, and can attach to algae and rocks. They are found in all marine waters, from surface tropical waters and shallow tide pools to the deep sea and polar regions. Most chaetognaths are transparent and are torpedo shaped, but some deep-sea species are orange. They range in size from 2 to 120 millimetres (0.1 to 4.7 in).

There are more than 120 modern species assigned to over 20 genera. Despite the limited diversity of species, the number of individuals is large.

Arrow worms are usually considered a type of protostome that do not belong to either Ecdysozoa or Lophotrochozoa.

Anatomy

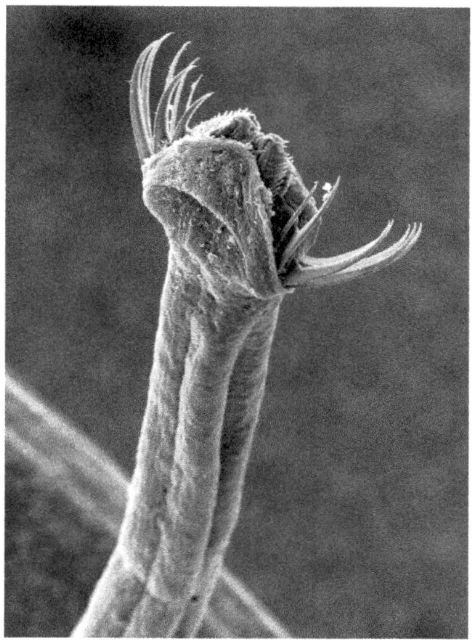

The jaw organ of a Chaetognath of the genus *Sagitta*

Chaetognaths are transparent or translucent dart-shaped animals covered by a cuticle. The body is divided into a distinct head, trunk, and tail. There are between four and fourteen hooked, grasping spines on each side of their head, flanking a hollow vestibule containing the mouth. The spines are used in hunting, and covered with a flexible hood arising from the neck region when the animal is swimming. All chaetognaths are carnivorous, preying on other planktonic animals.

The trunk bears one or two pairs of lateral fins incorporating structures superficially similar to the fin rays of fish, with which they are not homologous, however: unlike those of vertebrates, these are composed of a thickened basement membrane extending from the epidermis. An additional caudal fin covers the post-anal tail. Two chaetognath species, *Caecosagitta macrocephala* and *Eukrohnia fowleri*, have bioluminescent organs on their fins.

Chaetognaths swim in short bursts using a dorso-ventral undulating body motion, where their tail fin assists with propulsion and the body fins with stabilization and steering. Some species are known to use the neurotoxin tetrodotoxin to subdue prey.

The body cavity is lined by peritoneum, and therefore represents a true coelom, and is divided into one compartment on each side of the trunk, and additional compartments inside the head and tail, all separated completely by septa. Although they have a mouth with one or two rows of tiny teeth, compound eyes, and a nervous system, they have no respiratory or circulatory systems.

The mouth opens into a muscular pharynx, which contains glands to lubricate the passage of food. From here, a straight intestine runs the length of the trunk to an anus just in front of the tail. The intestine is the primary site of digestion and includes a pair of diverticula near the anterior end. Materials are moved about the body cavity by cilia. Waste materials are simply excreted through the skin and anus.

The nervous system is reasonably simple, consisting of a ganglionated nerve ring surrounding the pharynx. The dorsal ganglion is the largest, but nerves extend from all the ganglia along the length of the body. Chaetognaths have two compound eyes, each consisting of a number of pigment-cup ocelli fused together. In addition, there are a number of sensory bristles arranged in rows along the side of the body, where they probably perform a function similar to that of the lateral line in fish. An additional, curved, band of sensory bristles lies over the head and neck.

The arrow worm rhabdomeres are derived from microtubules 20 nm long and 50 nm wide, which in turn form conical bodies that contain granules and thread structures. The cone body is derived from a cilium.

Reproduction

All species are hermaphroditic, carrying both eggs and sperm. Each animal possesses a pair of testes within the tail, and a pair of ovaries in the posterior region of the main body cavity. Immature sperm are released from the testes to mature inside the cavity of the tail, and then swim through a short duct to a seminal vesicle where they are packaged into a spermatophore.(Sexually and Asexually)

During mating, each individual places a spermatophore onto the neck of its partner after rupture of the seminal vesicle. The sperm rapidly escape from the spermatophore and swim along the midline of the animal until they reach a pair of small pores just in front of the tail. These pores connect to the oviducts, into which the developed eggs have already passed from the ovaries, and it is here that fertilisation takes place.

The eggs are planktonic, or attached to algae, and hatch into miniature versions of the adult, without a well-defined larval stage.

Classification

Chaetognaths are traditionally classed as deuterostomes by embryologists. Lynn Margulis and K. V. Schwartz place chaetognaths in the deuterostomes in their *Five Kingdom* classification. Molecular phylogenists, however, consider them to be protostomes. Thomas Cavalier-Smith places them in the protostomes in his *Six Kingdom* classification. The similarities between chaetognaths and nematodes mentioned above may support the protostome thesis—in fact, chaetognaths are sometimes regarded as a basal ecdysozoan or lophotrochozoan. Chaetognatha appears close to the base of the protostome tree in most studies of their molecular phylogeny. This may explain their deuterostome embryonic characters. If chaetognaths branched off from the protostomes before they evolved their distinctive protostome embryonic characters, they might have retained deuterostome characters inherited from early bilaterian ancestors. Thus chaetognaths may be a useful model for the ancestral bilaterian. Studies of arrow worms' nervous systems suggests they should be placed within the protostomes.

Fossil Record

Due to their soft bodies, chaetognaths fossilize poorly. Even so, several fossil chaetognath species have been described. Chaetognaths appear to have originated in the Cambrian Period. Complete body fossils have been formally described from the Lower Cambrian Maotianshan shales of Yunnan, China (*Eognathacantha ercainella* Chen & Huang and *Protosagitta spinosa* Hu) and the Middle Cambrian Burgess Shale of British Columbia (*Oesia disjuncta* Walcott), a view challenged by Conway Morris (2009). A more recent chaetognath, *Paucijaculum samamithion* Schram, has been described from the Mazon Creek biota from the Pennsylvanian of Illinois. Chaetognaths were thought possibly to be related to some of the animals grouped with the conodonts. The conodonts themselves, however, are thought to be related to the vertebrates. It is now thought that protoconodont elements (e.g., *Protohertzina anabarica* Missarzhevsky, 1973), are probably grasping spines of chaetognaths rather than teeth of conodonts. Previously chaetognaths in the Early Cambrian were only suspected from these protoconodont elements, but the more recent discoveries of body fossils have confirmed their presence then.

References

- Sinigaglia, C.; et al. (2013). "The bilaterian head patterning gene six3/6 controls aboral domain development in a cnidarian". PLoS Biology. 11 (2): e1001488. PMC 3586664 . PMID 23483856. doi:10.1371/journal.pbio.1001488
- Lavelle, P. (July 1996). "Diversity of Soil Fauna and Ecosystem Function" (PDF). Biology International. 33. Retrieved 2009-04-20
- Butterfield, Nicholas J. (December 2006). "Hooking some stem-group 'worms': fossil lophotrochozoans in the Burgess Shale". BioEssays. Hoboken, NJ: John Wiley & Sons. 28 (12): 1161–1166. ISSN 0265-9247. PMID 17120226. doi:10.1002/bies.20507
- "Reading trees: A quick review". University of California Museum of Paleontology. Archived from the original on 15 April 2009. Retrieved 2009-04-13
- Budd, G.E.; Jensen, S. (May 2000). "A critical reappraisal of the fossil record of the bilaterian phyla". Biological Reviews. 75 (2): 253–295. PMID 10881389. doi:10.1111/j.1469-185X.1999.tb00046.x
- Ponder, W.F.; Lindberg, D.R., eds. (2008). Phylogeny and Evolution of the Mollusca. Berkeley: University of California Press. p. 481. ISBN 978-0-520-25092-5
- Gewin, V (2005). "Functional genomics thickens the biological plot". PLOS Biology. 3 (6): e219. PMC 1149496 . PMID 15941356. doi:10.1371/journal.pbio.0030219
- Halanych, K.M (2004). "The new view of animal phylogeny" (PDF). Annual Review of Ecology, Evolution, and Systematics. 35: 229–256. doi:10.1146/annurev.ecolsys.35.112202.130124. Retrieved 2009-04-17
- Kathryn Phillips (15 June 2008). "Living Fossil Memories" (PDF). Inside JEB. 211 (12): iii. doi:10.1242/jeb.020370
- Karleskint G, Richard Turner R and , James Small J (2012) Introduction to Marine Biology Cengage Learning, edition 4, page 445. ISBN 9781133364467
- Carlson, S.J (November 2001). "Ghosts of the Past, Present, and Future in Brachiopod Systematics". Journal of Paleontology. 75 (6): 1109–1118. doi:10.1666/0022-3360(2001)075<1109:GOTPPA>2.0.CO;2
- Lemaire, P (2011). "Evolutionary crossroads in developmental biology: the tunicates". Development. 138 (11): 2143–2152. PMID 21558365. doi:10.1242/dev.048975
- Platt HM (1994). "foreword". In Lorenzen S, Lorenzen SA. The phylogenetic systematics of freeliving nematodes. London: The Ray Society. ISBN 0-903874-22-9

- Ushatinskaya, G.T (2008). "Origin and dispersal of the earliest brachiopods". Paleontological Journal. Springer Verlag. 42 (8): 776–791. doi:10.1134/S0031030108080029
- Puce, S. (2007). "Symbiotic relationships between hydroids and bryozoans". International Symbiosis Society Congress Number 5. International Symbiosis Society. 44 (1–3): 137–143. Retrieved 2009-08-18
- Rouse, G.W. (2002). "Annelida (Segmented Worms)". Encyclopedia of Life Sciences. John Wiley & Sons, Ltd. doi:10.1038/npg.els.0001599
- Healy, J.M. (2001). "The Mollusca". In Anderson, D.T. Invertebrate Zoology (2 ed.). Oxford University Press. pp. 120–171. ISBN 0-19-551368-1
- Taylor, P.D.; Berning, B.; Wilson, M.A. (November 2013). "Reinterpretation of the Cambrian 'bryozoan' Pywackia as an octocoral". Journal of Paleontology. 87 (6): 984–990. doi:10.1666/13-029
- Paxton, H. (June 2005). "Molting polychaete jaws—ecdysozoans are not the only molting animals". Evolution & Development. 7 (4): 337–340. PMID 15982370. doi:10.1111/j.1525-142X.2005.05039.x
- Rouse, G. (1998). "The Annelida and their close relatives". In Anderson, D.T. Invertebrate Zoology. Oxford University Press. pp. 179–183. ISBN 0-19-551368-1
- Taylor, Paul D (October 2008). "Taxonomy of the bryozoan genera Oncousoecia, Microeciella and Eurystrotos". Journal of Natural History. 42 (39–40): 2557–2574. doi:10.1080/00222930802277640

PERMISSIONS

All chapters in this book are published with permission under the Creative Commons Attribution Share Alike License or equivalent. Every chapter published in this book has been scrutinized by our experts. Their significance has been extensively debated. The topics covered herein carry significant information for a comprehensive understanding. They may even be implemented as practical applications or may be referred to as a beginning point for further studies.

We would like to thank the editorial team for lending their expertise to make the book truly unique. They have played a crucial role in the development of this book. Without their invaluable contributions this book wouldn't have been possible. They have made vital efforts to compile up to date information on the varied aspects of this subject to make this book a valuable addition to the collection of many professionals and students.

This book was conceptualized with the vision of imparting up-to-date and integrated information in this field. To ensure the same, a matchless editorial board was set up. Every individual on the board went through rigorous rounds of assessment to prove their worth. After which they invested a large part of their time researching and compiling the most relevant data for our readers.

The editorial board has been involved in producing this book since its inception. They have spent rigorous hours researching and exploring the diverse topics which have resulted in the successful publishing of this book. They have passed on their knowledge of decades through this book. To expedite this challenging task, the publisher supported the team at every step. A small team of assistant editors was also appointed to further simplify the editing procedure and attain best results for the readers.

Apart from the editorial board, the designing team has also invested a significant amount of their time in understanding the subject and creating the most relevant covers. They scrutinized every image to scout for the most suitable representation of the subject and create an appropriate cover for the book.

The publishing team has been an ardent support to the editorial, designing and production team. Their endless efforts to recruit the best for this project, has resulted in the accomplishment of this book. They are a veteran in the field of academics and their pool of knowledge is as vast as their experience in printing. Their expertise and guidance has proved useful at every step. Their uncompromising quality standards have made this book an exceptional effort. Their encouragement from time to time has been an inspiration for everyone.

The publisher and the editorial board hope that this book will prove to be a valuable piece of knowledge for students, practitioners and scholars across the globe.

Index

A

Abiotic Characteristics, 5
Acidification, 7, 16
Amynthas Agrestis, 201
Aquarium Snails, 101
Aquatic Animal, 68
Aquatic Ecosystem, 1, 4, 118
Aquatic Mammal, 69, 104, 109
Aquatic Plant, 10, 41, 43, 49-50 61, 64
Asexual Reproduction, 41, 50, 199, 231-232
Autotrophic Organisms, 6
Avicularia, 220-221

B

Bacopa, 52-54, 67
Baikal Seal, 105, 107, 117, 138-141
Benthic Invertebrates, 13, 144
Biotic Characteristics, 5-6
Brachiopod, 160, 186, 204-207 209-215, 226, 228, 230, 237
Bryozoa, 160, 162, 164, 204 217, 219, 223-224, 226-229 232
Bycatch, 80, 105, 112, 120, 134, 143, 154

C

Caenogastropoda, 96
Canadia, 202
Cavefish, 68-73, 158
Ceratophyllum, 41-42, 51-52
Chaetognatha, 160, 164, 168, 229, 234, 236
Cloudina, 161, 183, 186, 189-190, 202
Colony Forms, 221-222
Community Patterns, 14

E

Echinoderms, 2, 161, 168-169, 179, 182, 184, 194, 205, 212 215-216, 228, 233
Equisetum Fluviatile, 41, 43, 65
Eutrophication, 3, 7, 16, 28, 59
Evolutionary Family Tree, 193, 226-227

F

Feeding Zooids, 217, 219
Fossil Record, 122, 173-174 176, 178, 180, 182, 184-189 202, 213-214, 218, 224, 226 237
Freshwater Ecosystem, 2, 6-7
Freshwater Ecosystems, 1-2, 6, 17
Freshwater Fauna, 7, 170
Freshwater Marshes, 36-37
Freshwater Snail, 68, 94, 100, 102

H

Halophyte, 39
Heterobranchia, 98-99
Heterotrophic Organisms, 6
Hippuris Vulgaris, 41, 43, 65

I

Intertidal Zones, 17, 24, 213
Invasive Species, 5, 7, 17-18, 43, 118
Iris Pseudacorus, 65
Isoetid, 42, 66

K

Kelp Forests, 28-29, 124, 128, 218
Keystone Species, 7, 114, 127

L

Lake Creation, 15
Lake Ecosystem, 7
Latitudinal Patterns, 15
Lentic Ecosystems, 6-8
Lentic Food Webs, 13-14
Lentic Waters, 7, 10, 16
Lophophore, 165, 206-211, 216-220, 228-229
Lophotrochozoans, 216-217, 237

M

Macrophyte, 7, 12, 64-67
Mangrove, 2, 18, 23-24, 27, 39
Marine Arthropods, 171
Marine Cnidarians, 166
Marine Coasts, 22
Marine Ecosystem, 17, 20, 128
Marine Habitats, 2, 18-22, 24, 76, 111, 160
Marine Invertebrates, 21, 160-161, 164, 168
Marine Molluscs, 169
Marine Topography, 21
Marine Worms, 32, 163-165, 167-168, 234

Marsh, 3-4, 34-39, 43, 64, 109
Mudflats, 26
Mysticetes, 123, 144-147, 151-152

N
Neritimorpha, 95

O
Ocean Currents, 18-21, 32
Ocean Traffic, 112, 120
Odontocetes, 123, 125, 143-146, 148-149, 151
Open Ocean, 18, 22, 31, 33, 124, 142
Organismal Respiration, 9

P
Parasitology, 101
Phragmochaeta, 202
Playa Lakes, 38
Pomacea Bridgesii, 96, 101-102
Potamogeton Perfoliatus, 61-64
Prairie Potholes, 38
Primary Producers, 9-10, 12-14, 16, 20, 28

R
River Dolphin, 68, 73-93, 104-106, 115, 118, 159
River Ecosystems, 4
Riverine Wetlands, 38
Rocky Shores, 25-26, 124

S
Sagittaria, 41, 44-47, 65, 67
Salt Marshes, 2, 17-18, 27, 36-37
Sandy Shores, 24-26
Seagrass, 18, 24, 29-30, 41, 54-55, 58-61, 67, 127
Seagrass Meadows, 18, 29-30, 58-61
Seamounts, 22, 33-34
Seeps, 19, 34, 129, 152, 199, 231
Species Richness, 13-14, 71, 123-124
Subspecies, 62, 73-75, 79, 82-84, 89, 102, 106, 138

T
Taxonomy, 7, 52, 55, 62, 66-67 72-73, 83, 92-94, 106, 120 122, 143, 211, 223-224, 238
Therapeutic Agents, 65
Trenches, 22, 34, 168, 217, 231

V
Vents, 2, 6, 19, 34, 129, 152, 168, 192, 199
Vernal Pools, 4, 38

W
Watershed Health Indicators, 65
Wet Meadows, 37
Wetlands, 1, 3-7, 16, 26, 28, 35-36, 38-39, 41, 66
Whale Fall, 152
Whale Pump, 129, 152
Wiwaxia, 175, 202